# Genetics and Ecotoxicology

# Current Topics in Ecotoxicology and Environmental Chemistry

Series Editors

Gary Rand, Peter Calow, and Michael A. Lewis

**Genetics and Ecotoxicology**
*Valery E. Forbes, editor, 1998*

*Current Topics in Ecotoxicology
and Environmental Chemistry*

# Genetics and Ecotoxicology

Edited by

**Valery E. Forbes, Ph.D.**
Department of Life Sciences and Chemistry
Roskilde University
Roskilde, Denmark

| USA | Publishing Office: | Taylor & Francis, Inc.<br>325 Chestnut Street<br>Philadelphia, PA 19106<br>Tel: (215) 625-8900<br>Fax: (215) 625-2940 |
| | Distribution Center: | Taylor & Francis, Inc.<br>47 Runway Road, Suite G<br>Levittown, PA 19057-4700<br>Tel: (215) 269-0400<br>Fax: (215) 269-0363 |
| UK | | Taylor & Francis Ltd.<br>1 Gunpowder Square<br>London EC4A 3DE<br>Tel: 171 583 0490<br>Fax: 171 583 0581 |

**GENETICS AND ECOTOXICOLOGY**

1 2 3 4 5 6 7 8 9 0

Printed by Braun-Brumfield, Ann Arbor, MI, 1998

A CIP catalog record for this book is available from the British Library.
&#8855; The paper in this publication meets the requirements of the ANSI Standard Z39.48-1984 (Permanence of Paper)

**Library of Congress Cataloging-in-Publication Data**

Genetics and ecotoxicology / [edited by] Valery E. Forbes
        p.  cm. -- (Current topics in ecotoxicology and environmental chemistry)
    Includes bibliographical references and index.
    ISBN 1-56032-715-4 (cloth: alk. paper)
    1. Environmental toxicology.  2. Ecological genetics.  I. Forbes, V. E. (Valery E.)
II. Series.
    RA1226.G46  1998                                                    98-28322
    571.9'5--dc21                                                       CIP

    ISBN 1-56032-715-4 (case)

# Contents

# Contributing Authors

**Donald J. Baird, Ph.D.,** *Institute of Aquaculture, University of Stirling, Stirling FK9 4LA, Scotland*

**Carlos Barata, Ph.D.,** *Institute of Aquaculture, University of Stirling, Stirling FK9 4LA, Scotland*

**Robert A. Browne, Ph.D.,** *Department of Biology, Wake Forest University, Winston-Salem, NC 27109, USA*

**Maria Jose Carmona, Ph.D.,** *Department de Microbiologia i Ecologia, Universitat de Valencia, Burjassot (Valencia), Spain*

**Michael H. Depledge, Ph.D.,** *Plymouth Environmental Research Centre, University of Plymouth, Drake Circus, Plymouth PL4 8AA, UK*

**Valery E. Forbes, Ph.D.,** *Department of Life Sciences and Chemistry, Roskilde University, PO Box 260, 4000 Roskilde, Denmark*

**Robert B. Gillespie, Ph.D.,** *Department of Biological Sciences, Indiana University-Purdue University, Fort Wayne, IN 46805-1499, USA*

**Dick Groenendijk, M.Sc.,** *University of Amsterdam, Department of Aquatic Ecology and Ecotoxicology, Kruislaan 320, 1098 SM Amsterdam, The Netherlands*

**Sheldon I. Guttman, Ph.D.,** *Department of Zoology, Miami University, Oxford, OH 45056, USA*

**Paul L. Klerks, Ph.D.,** *Department of Biology, The University of Southern Louisiana, Lafayette, LA 70504-2451, USA*

**Nicolay Mugue, Ph.D.,** *Institute for Developmental Biology, Moscow, Russia*

**Vibeke Møller, Ph.D.,** *Office of Pesticides, Danish Environmental Protection Agency, 1401 Copenhagen K, Denmark*

**Jaap Folkert Postma, Ph.D.,** *AquaSense Consultants, PO Box 95125, 1090 HC Amsterdam, The Netherlands*

**Lee R. Shugart, Ph.D.,** *LR Shugart and Associates, Inc., P.O. Box 5564 Oak Ridge, TN 37831-5564, USA*

**Manuel Serra, Ph.D.,** *Department de Microbiologia i Ecologia, Universitat de Valencia, Burjassot (Valencia), Spain*

**A. Jonathan Shaw, Ph.D.,** *Department of Botany, Duke University, Durham, NC 27708, USA*

**Terry W. Snell, Ph.D.,** *School of Biology, Georgia Institute of Technology, Atlanta, GA 30332-0230, USA*

**Christopher W. Theodorakis, Ph.D.,** *Department of Wildlife and Fisheries Sciences, Texas A & M University, College Station, TX 77801, USA*

**Judith S. Weis, Ph.D.,** *Department of Biological Sciences, Rutgers University, Newark, NJ 07102, USA*

**Peddrick Weis, DDS.,** *Department of Anatomy, Cell Biology and Injury Science, New Jersey Medical School, UMDNJ, Newark, NJ 07103, USA*

# Foreword

Biology is self-evidently multidisciplinary but rarely interdisciplinary. By this I mean that biological systems always involve a rich blend of physical, chemical, structural, and informational processes, as well as many others, and yet partly because of the way we are educated, biologists rarely manage to step outside particular disciplines to attempt to come to terms with biological systems and the problems they face.

Very fundamentally, though, the fact that organisms have genotype and phenotype means that they broker the interaction between information systems based on a precise molecular organization and biochemical, physiological, and developmental processes in responding to their environments. The way that the genotype controls the phenotype and the way that it is shuffled and distributed by reproductive processes plays a big part in determining the responses of individuals, the distribution of response types between individuals, and the change in these distributions through time.

Therefore, any understanding of the way toxic chemicals affect ecological level systems and processes—the stuff of ecotoxicology—has to account for these interfaces. Yet ecotoxicology most often concentrates on phenotypes, the survival and reproductive responses of a few individuals, ignoring the genetics behind what they are and do, and hence obtaining only a partial understanding of the ecological response. This interdisciplinary volume makes an explicit attempt to put this right. Ranging from molecular to classical genetics, from plant to animal, from asexual to sexual systems, it touches on some very fundamental issues of evolutionary biology, e.g., why mode of reproduction matters, but most importantly for this series, why it is important to take genetics into account in understanding if and how chemical contaminants impact populations.

*Peter Calow*

# Preface

This book marks the first volume in Taylor and Francis's series on *Current Topics in Ecotoxicology and Environmental Chemistry*. The volumes will be designed to cover advances and timely reviews in these fields. This particular volume was inspired partially by my own interests in determining the causes of variability among organisms in their responses to toxicant exposure. Such differences may be critical for understanding the mechanistic basis of responses to toxicants and are also of practical relevance for ecotoxicological testing and risk assessment. In addition, there is growing concern that the genetic consequences of exposure to toxic chemicals may have profound implications for the long-term persistence of populations and species.

It seems clear that ecotoxicology can benefit from a more explicit incorporation of genetics and that there is a pressing need for research at the interface of these two important fields of study. Authors were selected that could make a significant contribution to both basic and applied issues. Thus, the topics covered here include toxicant-caused genetic change (e.g., selection for resistant genotypes, allozyme frequency changes, loss of genetic diversity), toxicant-caused genetic damage (e.g., DNA adducts, strand breakage), the role of genetic diversity in controlling organism responses to toxicants, and consideration of the extent to which genetic variability may be of concern for ecological risk assessment.

The book is primarily designed for active researchers and others involved in ecotoxicology at a professional level. However, it could also make a valuable contribution to a post-graduate course, and many of the issues covered should be of interest to environmental managers and those involved in regulatory practice as well.

I am extremely grateful to the contributors to this volume for sharing their insights into the exciting interface of genetics and ecotoxicology and for tolerating my nagging during the various stages of manuscript preparation. I am also grateful to the reviewers for their prompt and thoughtful comments. Thanks also to the series editors for their support for this project and to the editorial staff at Taylor & Francis for all of their hard work and attention to detail.

*Valery Forbes*

*Genetics and Ecotoxicology*
Edited by V. E. Forbes
Copyright © 1999 Taylor & Francis

# 1

# Genetics and Ecotoxicology—Insights from the Interface

Valery E. Forbes

**Abstract.** The ten contributions to this volume address a number of key issues that, taken together, summarize our current understanding of the relationship between genetics and ecotoxicology. They do so from very different perspectives—they examine different taxonomic groups and contaminant classes, use different methodological techniques, and approach their questions using a variety of theoretical and empirical approaches. What emerges from an analysis of these contributions are some important, and a few surprising, conclusions that bear on the consequences of toxicant exposure for genetic variability, the importance of genetic variation in determining organism responses to toxicants, and the extent to which the study of genetic variation has a role to play in ecological risk assessment. In addition, gaps in our present understanding of genetic and ecotoxicological processes have been identified that suggest important challenges for future work in these fields.

**Keywords.** Adaptation; genetic variability; molecular damage; risk assessment.

## IMPLICATIONS OF EXPOSURE TO TOXICANTS FOR GENETIC VARIABILITY

Much of the literature dealing with the potential effects of toxicant exposure on the genetic variability of natural populations has been based on two major concerns. The first concern is that exposure to toxicants may cause reductions in the amount of genetic variability in populations, which will lead to a loss of adaptability and hence increase the probability of population extinction. The second concern is that exposure to toxicants may select for resistant genotypes and that, possibly as a result of various costs associated with tolerance, such genotypes have reduced fitness (relative to nonresistant genotypes) in the absence of exposure. Clearly, if either or both of these concerns are justified, the long-term ecological and evolutionary implications are serious. It is therefore critical to evaluate the extent to which empirical evidence has been produced in their support.

As summarized by Gillespie and Guttman (chapter 4), many studies have reported reductions in allozyme variation in populations from contaminated habitats. But as these authors point out, changes in allozyme frequencies (even if selected by toxicant exposure) do not necessarily indicate negative consequences for the population at hand. They argue that only if the changes in allozyme frequencies are associated with reduced average heterozygosity or with reduced fitness should the allozyme changes be considered negative. In their studies they were able to demonstrate a link between allozyme frequency and tolerance to pollutants. However, the evidence that such allozyme changes were associated with negative impacts on populations is less clear.

Snell et al. (chapter 9) use simulation models to directly address the issue of genetic diversity loss via selection and drift at nonselected loci in rotifer populations. They found that heterozygosity loss due to toxicant-caused reductions in population growth rate were minimal. Rather than loss of genetic diversity per se, the most serious consequence of reduced population growth rate was a reduction in resting egg production. Resting eggs, which are produced sexually, may remain dormant for several years under environmental conditions that would be lethal to adults. Thus, the production of resting eggs by cyclical parthenogens such as rotifers is crucial for allowing populations to become reestablished after periods of environmental stress. Also, it appears that sexual reproduction is inhibited at lower toxicant concentrations than is asexual reproduction in these animals.

Negative effects associated with the adaptation of chironomid populations to cadmium included increased mortality and larval development time in clean conditions (Postma and Groenendijk, chapter 5). However, gene flow from adjacent nonadapted populations had an important influence on the evolution of tolerance in exposed populations and contributed to maintaining their genetic diversity.

Selection against tolerant plant genotypes has been shown to occur on uncontaminated soils, consistent with the idea that there are costs associated with tolerance (Shaw, chapter 2). However, as Shaw points out, the evidence for costs of tolerance is indirect, and the exact nature of such costs remains poorly understood. Again, gene flow between tolerant populations and adjacent nontolerant populations can have an important influence on population genetic structure and diversity.

In chapter 3 (Weis et al.), a single species–toxicant combination, *Fundulus heteroclitus* and mercury, is used to demonstrate the potential complexities that may provide difficulties in identifying the existence and genetic basis of tolerance. For example, the authors found that fish embryos from an exposed site exhibited increased tolerance to mercury, whereas larvae and adults from this site did not. These authors and others have performed extensive allozyme analyses of this species along the Atlantic coast of the United States, and although different subspecies were identified, genetic markers clearly associated with pollutants and enhanced tolerance could not be detected. Nor were there indications from these extensive studies that tolerant populations are less genetically diverse than their nontolerant conspecifics.

At the community level, severely polluted habitats are often less species rich and in this sense have a lower genetic diversity than unpolluted habitats. However, the evidence that toxicant exposure reduces genetic variability at the population level and is accompanied by reductions in the adaptability of toxicant-resistant populations is weak. Negative effects on fitness-related traits have been observed in tolerant populations in some cases (e.g., Shaw, chapter 2; Gillespie and Guttman, chapter 4; Postma and Groenendijk, chapter 5), but not others (e.g., Shaw, chapter 2; Gillespie and

Guttman, chapter 4). Evidence that toxicants select for particular life-history traits, such as earlier age at maturity (e.g., Weis et al., chapter 3) can have positive effects on population growth rate and could provide a mechanism for the spread of tolerant genotypes into adjacent uncontaminated habitats.

Much of the evidence for selection against tolerant genotypes derives from cases in which exposures have been extreme. For species in which tolerant genotypes have evolved—and even in those cases in which selection against tolerant genotypes in adjacent uncontaminated sites has been demonstrated—there are no cases in which the evolution of tolerance has resulted in the decline of the species. Largely, this appears to be due to extensive gene flow from uncontaminated sites.

Although most concern has focused on decreases in genetic diversity as a result of toxicant exposure, Klerks (chapter 6) notes that exposure to certain toxicants may increase genetic diversity as a result of increased mutation rate. In addition, Gillespie and Guttman (chapter 4) point out that toxicant-caused reductions in population size can increase, as well as decrease, the frequency of rare alleles. Most of what we know about the consequences of toxicant exposure for the evolution of tolerance is based upon single toxicants acting in isolation, and the majority of studies have concerned either heavy metals or pesticides. Using a combination of theoretical and empirical approaches, Klerks considers how contaminant complexity is likely to influence the evolution of tolerance. He argues that exposure to a mixture of contaminants may reduce selection intensity for the individual toxicants, thereby reducing the likelihood of tolerance evolution. This argument is supported by studies of pesticide and antibiotic resistance in which combinations of independently acting toxicants have proved effective in slowing down the development of resistance in the target species. However, as Klerks points out, failing to detect tolerance in exposed populations can be due to a number of factors, and great care must be taken when discounting possible explanations. The fact that contamination of natural habitats frequently involves mixtures of substances emphasizes the importance of exploring whether the genetic consequences of exposure to complex mixtures differ from those occurring in response to single toxicants.

## IMPLICATIONS OF GENETIC VARIATION
## FOR RESPONSES TO TOXICANTS

It has long been recognized that the existence of genetic variation for tolerance is a prerequisite for the evolution of tolerance. Evidence for genetic variability in tolerance to toxic chemicals is widespread, as reviewed by a number of the authors in this volume. Less is known about the specific genetic basis of tolerance and the degree to which it is influenced by the level and duration of exposure. A long history of studies of plants growing in metal-contaminated habitats suggests that the presence of tolerance is controlled by a limited number of major genes in combination with modifier genes that influence the level of tolerance (Shaw, chapter 2). However, much of this work deals with contamination that has been abrupt and severe, and it may be that gradual, low-level contamination results in a different set of genetic consequences.

Genetic variation has posed difficulties for ecotoxicologists in that it has been associated with substantial differences in the phenotypic responses of different populations to toxicant exposure. Such variability in toxicant responses complicates attempts to extrapolate biological effect concentrations from standardized laboratory test results to nature. A main concern in this regard is that we may be under-protecting certain populations to the extent that sensitive genotypes are not accounted for. Not only do many of our standard laboratory tests employ populations having little or no genetic variability (e.g., clones), but there is evidence (discussed by Baird and Barata, chapter 11) that the genotypes used for standard tests may be more tolerant than average, possibly yielding biological effect concentrations that are even less protective than recognized.

Although genetic variation is generally believed to be important in determining the responses of populations to toxicants, the existence of genotype × environment interactions prevents broad generalizations as to the relative importance of genetics versus environment. Forbes et al. (chapter 10) found that genetic variability among clones of *Artemia* played a minor role in controlling the response of these organisms to copper exposure. In contrast, studies by Baird and colleagues have demonstrated that genetic differences in tolerance to toxicant exposure can, in some cases, be substantial and result in over three order of magnitude differences in $LC_{50}$ values among clones of a single species. This body of work provides an excellent example of the importance of genotype × environment interaction (see especially Figs. 2 and 4 in chapter 11) and highlights the challenges we face in trying to extrapolate ecotoxicological responses across genotypes and/or toxicants.

## WHAT ROLES DOES GENETICS HAVE TO PLAY
## IN ECOLOGICAL RISK ASSESSMENT?

Shugart's contribution (chapter 8) describes how specific types of damage to the structure of DNA can provide valuable endpoints of toxicant exposure. The usefulness of such molecular biomarkers for ecological risk assessment has been questioned on the grounds that DNA damage may not be coupled to ecological effects. As Shugart notes, the relevance of genetic damage for predicting effects at population and higher levels is problematic. However, as indicators of exposure, changes in DNA can be quite powerful tools that, in combination with independently derived ecological effect levels, can play an important part in the risk assessment process. Advantages of DNA markers as indicators of exposure over traditional measures of environmental exposure concentrations include the fact that toxicant bioavailability is accounted for, the techniques are sensitive, and they do not necessarily require that monitoring organisms be destroyed.

Tests of survival and reproduction in the water flea, *Daphnia magna*, provide key endpoints for ecological effects assessment and are among the few ecotoxicological tests that provide the basis for legislative decisions. Present internationally agreed test protocols (OECD 1997) are confined to examining survival and asexual reproduction in single clones of *Daphnia*, despite the fact that, in nature, *Daphnia* reproduces both sexually and asexually, and its populations are made up of a mixture of genotypes. As discussed by Baird and Barata (chapter 11) and Forbes et al. (chapter 10) parthenogenetically reproducing clones offer several advantages, the most important of which is that they may facilitate identification and quantification of the role of genetic factors in controlling responses to toxicants. However, the experiments of Snell et al. (chapter 9 and references therein), using another cyclical parthenogen, indicate that there may be significant differences in the exposure concentrations at which sexual and asexual reproduction become impaired. If the pattern found by Snell et al. turns out to apply to other cyclical parthenogens such as *Daphnia*, then confining test protocols to the asexual phase of the life-cycle of test populations is likely to underestimate toxicant risks to natural populations. Despite their obvious convenience, parthenogentically reproducing organisms may turn out to be a bad choice for ecotoxicological effects assessment.

If, as Baird and Barata (chapter 11) suggest, the degree of genetic variation in organism response to toxicant exposure is substance and species specific, how can we deal with this uncertainty in ecotoxicological risk assessment? As one possibility, Baird and Barata suggest that the distribution of sensitivities of genotypes in selected populations be examined. The distribution of genotype sensitivities could be used to estimate species protection criteria in the same way that species sensitivity distributions have been extrapolated to derive ecosystem protection criteria (Stephan et al. 1985; Wagner and Løkke 1991; Aldenberg and Slob 1993). The extrapolation models applied at the species-to-ecosystem level suffer from a number of weaknesses (Forbes and Forbes 1994). Among them are that the species used to fit the sensitivity distributions are neither a random sample of species nor necessarily even members of the ecosystem(s) for which extrapolations are being made. Fitting a sensitivity distribution at the genotype-to-population level could at least avoid these two potentially important pitfalls and may help to better quantify the extent to which uncertainty due to genetic influences needs to be incorporated into risk estimates.

Another approach for incorporating genetic uncertainty (also discussed by Baird and Barata, chapter 11) could be to use the observed clone-to-clone differences as an argument for maintaining or even increasing the application factors for prediction of no-effect concentrations from test data. In this regard, it is tempting to conclude that if different genotypes of a single population vary substantially from each other in their responses to toxicant exposure, such variability will be magnified if we compare different populations and species. However, the results presented by Forbes et al. (chapter 10) challenge this assumption. They found that geographically distinct populations of *Artemia* species responded similarly to copper exposure whereas the major source of variability was among individuals within single clones.

Quantifying uncertainty in susceptibility to toxicants is an important aspect of both ecological and human health risk assessment, and genetic factors can be an important source of this uncertainty. There is a pressing need for better models that can account for genetic differences in susceptibility to toxicant exposure and can incorporate this uncertainty into risk estimates.

## CHALLENGES FOR THE FUTURE

Although there are good reasons to believe that adaptation of organisms to toxicants involves certain costs (Forbes and Calow 1996; Theodorakis and Shugart chapter 7), identifying and quantifying these costs remain a challenge. Shaw (chapter 2) discusses the difficulties in distinguishing the costs of tolerance versus those associated with adaptations to other features of contaminated habitats. A recent study of copper tolerance in *Mimulus guttatus* demonstrated an absence of highly tolerant plants in the field, despite the ability to rapidly and artificially select for extreme levels of tolerance (Harper et al. 1997). Although such observations are consistent with the idea that tolerance is costly, the authors were unable to find direct evidence for costs of tolerance in this species.

Developing ecotoxicological tests to deal effectively with complex mixtures is an area of active study. With the possible exception of pesticides, very little work to date has considered the effects of mixtures of toxicants from a genetic perspective. Given that, in the field, exposed populations are more often than not exposed to mixtures of contaminants, it is critical that we devote research efforts to this area.

Continuing advances in molecular biological techniques increase our ability to determine whether populations are being exposed to potentially dangerous substances (Theodorakis and Shugart, chapter 7; Shugart, chapter 8). We may expect these techniques to receive wider usage in future environmental monitoring programs. In the past, opportunistic species have been considered to be poor choices for environmental monitoring because of their suspected insensitivity to contaminant impact. However, it may be precisely the genetic features of such species that allow them to persist in contaminated habitats that may make them valuable ecological indicators. Theodorakis and Shugart (chapter 7) argue that adaptation in opportunistic, and not specialist, species provides the mechanism for taxonomic simplification of communities in contaminated habitats. They suggest that an increase in the frequency of resistant genotypes in opportunistic species may provide a tool for determining whether other members of the community are at risk. For this approach to be advantageous, requires that changes in gene frequency in opportunists occur at lower exposure concentrations than those leading to the loss of sensitive, specialist species. The evidence for this assumption is limited, and further work involving a combination of molecular techniques and field community sampling is required to confirm it. As shown by Theodorakis and Shugart (chapter 7), tools such as RAPD can greatly facilitate the molecular side of this work.

A number of hypotheses have been forwarded to explain observed differences in the degree of genetic variation in response to different toxicants, yet none have so far been verified. The advantages of employing clonally reproducing species to test these hypotheses have been described (Forbes et al., chapter 10; Baird and Barata, chapter 11). In doing so, it is important to recognize that there may be important differences in the genetic composition of species that exhibit purely asexual reproduction versus those that experience cyclical bouts of sexual reproduction. Heterozygosity, ploidy, and genetic load are among those factors that are likely to differ. Whether they result in general differences in tolerance and, if so, whether the frequency of sex in cyclical species has a measurable effect on the genetic basis of tolerance are questions that remained to be addressed.

Gillespie and Guttman (chapter 4) highlight the importance of linking allozyme changes to measures of individual performance (e.g., survival, reproductive output, etc.) and population fitness. Indeed, to be useful for ecological risk assessment, it is essential that these links be extended at least to the population level. Recent applications of life-table response experiments to ecotoxicology (Kammenga et al. 1996; Levin et al. 1996; Hansen et al. in press; Sibly in press) indicate that the relationship between toxicant-caused impairment of life-history traits of individuals (e.g., survival, age at maturity, number of offspring, etc.) and population growth rate is not straightforward. Some traits may be quite sensitive to toxicant exposure, yet changes in them have little effect on population growth rate. The reverse may also occur. Further developments in both the theoretical models and empirical testing of their predictions under ecologically realistic conditions are important areas for future research.

As the chapters that follow demonstrate, the fields of genetics and ecotoxicology have much to offer each other. Our understanding of the relationship between genetic variability and the responses of populations to toxicants has advanced on many fronts. Yet many questions remain. Finding their answers will require a greater integration of genetics and ecotoxicology—a goal toward which this volume can hopefully provide valuable inspiration.

## REFERENCES

Aldenberg, T., and W. Slob. 1993. Confidence limits for hazardous concentrations based on logistically distributed NOEC data. *Ecotoxicology and Environmental Safety* 25:46–63.

Forbes, V. E., and P. Calow. 1996. Costs of living with contaminants: Implications for assessing low-level exposures. *Biological Effects of Low Level Exposures (BELLE)* 4:1–8.

Forbes, V. E., and T. L. Forbes. 1994. *Ecotoxicology in theory and practice*. London: Chapman and Hall.

Hansen, F. T., V. E. Forbes, and T. L. Forbes. In press. The effects of chronic exposure to 4-n-nonylphenol on life-history traits and population dynamics of the polychaete *Capitella* sp. 1. *Ecological Applications* (in press).

Harper, F. A., S. E. Smith, and M. R. Macnair. 1997. Where is the cost in copper tolerance in *Mimulus guttatus*? Testing the trade-off hypothesis. *Functional Ecology* 11:764–774.

Kammenga, J. E., M. Busschers, N. M. Van Straalen, P. C. Jepson, and J. Bakker. 1996. Stress induced fitness reduction is not determined by the most sensitive life-cycle trait. *Functional Ecology* 10:106–111.

Levin, L., H. Caswell, T. Bridges, C. Dibacco, D. Cabrera, and G. Plaia. 1996. Demographic responses of estuarine polychaetes to sewage, algal, and hydrocarbon additions: Life-table response experiments. *Ecological Applications* 6:1295–1313.

OECD. 1997. Guideline for testing of chemicals no. 202. *Daphnia* sp. acute immobilization and reproduction test. Paris: Organization for Economic Cooperation and Development.

Sibly, R. M. In press. Efficient experimental designs for studying stress and population density in animal populations. *Ecological Applications* (in press).

Stephan, C. E., D. I. Mount, D. J. Hanson, J. H. Gentile, G. A. Chapman, and W. A. Brungs. 1985. Guidelines for deriving numeric national water quality criteria for the protection of aquatic organisms and their uses. PB85-227049. Duluth, MN: US Environmental Protection Agency.

Wagner, C., and H. Løkke. 1991. Estimation of ecotoxicological protection levels from NOEC toxicity data. *Water Research* 25:1237–1242.

*Genetics and Ecotoxicology*
Edited by V. E. Forbes
Copyright © 1999 Taylor & Francis

# 2

# The Evolution of Heavy Metal Tolerance in Plants: Adaptations, Limits, and Costs

## A. Jonathan Shaw

**Abstract.** Habitats contaminated with heavy metals provide natural laboratories for studies of evolution in plant populations. Understanding the genetic basis of tolerance is crucial for subsequent studies of evolutionary processes. Recent work on several species and metals suggests that tolerance is effected by a limited number of major genes, with modifiers that condition levels of tolerance rather than the presence of tolerance per se. Natural selection on highly contaminated mine tailings is primarily stabilizing, whereas selection on plants in closely adjacent uncontaminated soils appears to be directional (against tolerant individuals). Selection can occur at multiple stages in the life cycle, including both the diploid sporophyte phase and haploid gametophytes. In mosses, selection occurs primarily on haploid gametophytes, which constitute the perennial stage of the life cycle. Theory and some empirical data suggest that selection among haploid pollen in the styles of flowering plants growing in contaminated habitats could be an important factor in tolerance evolution, but experimental work to date has not confirmed selection among pollen individuals. Rather, selection occurs among developing ovules in pistils enriched with metals. The general restriction of tolerant plants to contaminated habitats and steep clines in tolerance from mine tailings onto adjacent "normal" soils suggest that there are costs associated with tolerance and that tolerant individuals are at a disadvantage in the absence of metal contamination. The nature of such costs is poorly understood, however, and additional research is needed in this area.

**Keywords.** Directional selection; ecotypes flowering plants; genetic basis of tolerance; heavy metal tolerance; mosses; selection; stabilizing.

## INTRODUCTION

That some plants can evolve tolerance of substrate heavy metal contamination has been recognized for over 60 years (Prat 1934). Early work demonstrated that plants growing on contaminated substrates are often genetically distinct from conspecifics on normal soil, suggesting that tolerant races, or ecotypes (Turreson 1922), can evolve rapidly. The seminal review of metal tolerance in plants published by Antonovics et al. (1971) stimulated interest in contaminated habitats as natural laboratories for studying evolution. Over 50 species of plants, algae, and fungi are now known to have evolved tolerance to metals (Shaw 1990).

A great deal of work has been devoted to understanding the evolution of metal-tolerant plants. Nevertheless, many population-level processes are still poorly understood and tolerance evolution continues to provide a valuable model system for investigating genetic and ecological processes in natural populations. A number of excellent reviews of the subject have been published since the paper by Antonovics et al. (1971), and a general review will not be provided here (see Antonovics 1975; Baker 1987; Bradshaw 1984; Bradshaw and McNeilly 1981; Macnair 1987, 1993; Purvis and Halls 1996; Shaw 1990). Instead, this chapter will focus on evolutionary processes in populations subject to metal contamination and on the constraints and limits to tolerance evolution. The present review deals primarily with angiosperms and bryophytes, which have been studied most intensively.

Several generalities can be derived from studies on metal tolerance in plants. One is that some species evolve tolerance and some do not. Another is that, in general, tolerant plants of a species are restricted to sites that are contaminated. These observations suggest that there are genetic limitations on the evolution of tolerance such that only some species have the capacity for tolerance evolution, and that certain poorly understood constraints tend to restrict the gene(s) that confer tolerance to habitats where they are positively selected.

The significance of metal tolerance as a model system derives in large part from the fact that the predominant selective factor (toxic metals) is clear. Although contaminated habitats such as mine tailings often differ from "normal" sites in other environmental features (e.g., low nutrients and organic matter, extreme drought, poor soil texture, low overall plant density; Antonovics et al. 1971; Farago 1981), metal tolerance per se is necessary (though not sufficient) for colonization and persistence. In addition, many of the sites where tolerance evolution has been studied are of anthropogenic origin, and their ages can be estimated. Thus, the speed at which tolerant populations develop can be assessed, and areas differing in age and stage of the evolutionary process can sometimes be compared (e.g., Wu et al. 1975). Levels of contamination can be readily determined using standard chemical analyses so that patterns and degrees of plant tolerance can be related to varying habitat contamination.

## THE GENETIC BASIS OF TOLERANCE

Reviews of studies on the genetic basis of metal tolerances in wild plants have been provided by Macnair (1990, 1993). Early work showing that tolerance appears to vary continuously led to the interpretation that the trait is controlled by many genes, each with small effect (Antonovics et al. 1971). Macnair (1983, 1990) argued, however, that this interpretation is an artifact of using tolerance indices to assess levels of tolerance. Tolerance indices are calculated as the ratio of growth (typically root growth) in some specified metal treatment/growth in a control treatment. Macnair et al. have pointed out that tolerance indices are subject to high levels of statistical noise because the two component variables each typically have differently skewed distributions. Moreover, rate of root growth (in the absence of metals) is itself variable and may be genetically independent of tolerance per se (Humphries and Nicholls 1984). Discontinuous variation in tolerance can be obscured by such difficulties in measurement. For these reasons,

Macnair (1990) advocates estimating tolerance in terms of absolute root growth in medium supplied with a metal concentration that fully inhibits growth in nontolerant plants but permits growth in tolerants. Finding such a concentration, however, may be more difficult in some species than in others (Schat and Ten Bookum 1992).

Macnair (1983) crossed nontrue breeding copper tolerant plants of *Mimulus guttatus*, selfed the progeny, and found three types of resulting families: tolerant nonsegregating, nontolerant nonsegregating, and segregating. A 1:1:2 ratio of the three types suggests one major gene controlling tolerance. Tolerance appears to be dominant over nontolerance. Variation in levels of tolerance and minor departures from expected Mendelian ratios of tolerant and nontolerant offspring reported by Macnair (1983) and Macnair et al. (1993) might be explained by additional segregating genes that affect tolerance. Recent work by Smith and Macnair (1998) indicates that there is one (or more) modifier gene hypostatic to the major tolerance gene in the sense that the modifier is expressed only in tolerant plants and affects the level of tolerance (as opposed to the presence of tolerance per se). The evidence regarding additional modifiers is less clear.

Schat and Ten Bookum (1992) and Schat et al. (1993) likewise found that copper tolerance in *Silene vulgaris* is conditioned by one major gene plus up to three modifiers (probably two) that are completely hypostatic to the major gene. The most significant difference between the results of Schat and Ten Bookum (1992) on *Silene* and those of Macnair et al. on *Mimulus* is that a higher proportion of the variation in levels of tolerance is attributable to the modifiers in *Silene*. Schat et al. (1993) detected at least two different major genes that act additively for copper tolerance in *S. vulgaris*. One was found in only one out of three German populations studied, whereas the other was common to the other three populations. The same set of modifiers appeared to be present in all populations.

The genetic basis of tolerance to arsenic appears to follow a pattern similar to that revealed for copper. In *Agrostis capillaris*, arsenic tolerance appears to be controlled by one or a limited number of genes that may interact nonadditively in determining levels of tolerance (Watkins and Macnair 1991). In general, crossing data suggest that a single major gene effects tolerance in *Holcus lanatus*, but several crosses yielded an excess of tolerant offspring. Macnair et al. (1992) thought that variable penetrance of the tolerance major gene across families differing in genetic background best explained their observations. A gene for albinism cosegregates with arsenic tolerance in *H. lanatus*.

There are still very few cases in which the genetic basis of metal tolerance in wild plants is well understood, but the emerging picture is consistent with knowledge about the genetics of tolerance in agronomic plants (Aniol and Gustafson 1990; Macnair 1993). A limited number of major genes, rather than classic polygenic inheritance involving many genes each with small effect, is also more consistent with the observed specificity of tolerance to individual metals.

Lande (1983) argued that most adaptations are more likely to occur via the spread of minor than major genes, whereas Macnair (1991) presented a model in which adaptation, especially in situations like the evolution of metal tolerance, is more likely to occur by the spread of major genes having large phenotypic effects. The critical element is that only those variants substantially more tolerant of the metals than "wild type" individuals can survive, and such qualitative differences in tolerance are unlikely to be provided by a polygenic system. Once tolerance is achieved, increases in the levels of tolerance could result from selection in which extreme phenotypes enjoy a selective advantage. Hypostatic modifiers of the major tolerance gene provide a mechanism of response to selection for increased tolerance. If modifiers are completely hypostatic to the major tolerance gene, then under extreme selection on severely contaminated soils their spread should occur more readily than the spread of minor genes that confer low levels of tolerance in the absence of the major gene (Smith and Macnair 1998). Macnair's (1991) model applies especially to situations such as colonization of highly contaminated mine tailings, where selection is abrupt and severe. In habitats with chronic but lower levels of contamination, where tolerance need not be "all-or-nothing," a polygenic model may apply. Much less is known about the genetics of tolerance in plants subjected to lower levels of accumulated toxins.

Individuals with intermediate levels of tolerance between fully tolerant and nontolerant plants are frequently described in studies of natural populations. Some of the intermediacy, no doubt, reflects the inaccuracy of measuring tolerance experimentally, but additional factors may also be involved. Variable penetrance of tolerance genes in different genetic backgrounds and additional modifiers that are not completely hypostatic to the major genes could contribute to variation in tolerance levels. Phenotypic expression of dominance relationships can also vary with the environment in which tolerance is measured. Previous work on the genetics of tolerance, even on just a few species sampled from a limited number of sites, has shown that more than one tolerance gene may be present within species, and that these genes may be expressed additively or nonadditively if present in the same plant. It seems likely that additional genes will be found, and that major and minor genes for tolerance, even if relatively limited in number, could contribute to variation in tolerance found among conspecific plants and populations. Nicholls and McNeilly (1982) found that in *A. capillaris*, three types of response to a series of copper concentrations could be detected in a sample of plants from seven different copper mines. These authors used root growth (not tolerance indices) to measure tolerance, and interpreted the variation they observed as resulting from independent evolution of tolerance within the species. Their results are consistent with different combinations of genes having been selected at different sites.

Although the number of tolerance genes thus far detected in *M. guttatus, S. vulgaris*, and *H. lanatus* are limited, sampling from the range of these species has not been extensive. Most of the plants utilized in crossing studies involving *M. guttatus*, for example, are ultimately derived from a very limited number of plants collected at two California sites. Macnair (1981) was not able to detect any tolerant plants from a British population of *M. guttatus*, reflecting stochastic variation in the distribution of genes for tolerance (Macnair and Watkins 1983). It thus seems unlikely that the full story relating to the number, occurrence, or frequency of tolerance genes is yet available for any species.

## NATURAL SELECTION IN METAL CONTAMINATED HABITATS

### The Time Scale for Tolerance Evolution

Crude estimates can often be made concerning the time scale for tolerance evolution. Substrates naturally enriched with one or more metals have, of course, existed for millions of years. Mineralized strata outcrop worldwide and are especially abundant in many mountainous areas because of the metamorphism associated with mountain formation. Metaliferous outcrops are especially abundant and extensive in south-central Africa and have been available to plants for millions of years. Indeed, Brooks and Malaisse (1985) suggest that mineralized outcrops in the Shaban Copper Arc and the Zambian Copperbelt have been available to plants since the origin of angiosperms during the Mesozoic Era. The significance of such habitats for plant evolution is amply demonstrated by the high percentage of endemics on metal-enriched substrates (Brooks and Malaisse 1985). In Shaba, at least 50 angiosperm taxa are endemic to these sites (>15% of the metallophyte flora). Disruptive selection associated with metal-enriched substrates clearly leads to evolutionary divergence in many species. Over long periods of time, such evolutionary processes can result in the origin of endemic taxa.

Mine wastes are of more recent origin. The Welsh copper and lead/zinc mines at Drws-y-Coed and Trelogan, respectively, were first worked in the 13th century, and present tailings with tolerant plants probably date from the 19th century (McNeilly and Antonovics 1968). Mine tailings in the Piedmont of North Carolina date from the early 19th century when some of the lead/zinc mines were important sources of ores for the manufacture of confederate ammunition (Wickland 1983). Compared to the African outcrops, a much lower number of endemic metallophyte taxa is known from mine wastes in North America and Europe, where the ages of contaminated sites are measured in decades to centuries, rather than in (tens of) millions of years. Metal tolerance was probably selected in some mining areas prior to human activity, where ores were exposed naturally at the surface. However, the low frequency of endemic taxa on metaliferous substrates in North America and Europe suggests either that large scale speciation has not occurred, perhaps due to limited sizes of enriched exposures, or that endemic taxa have gone extinct because of climate change, including glaciation. Populations growing on mine tailings, from which most of our information about evolutionary processes involved in tolerance evolution has been derived, are relatively recent in origin. Patterns observed in mine populations are no older than decades to a century or two.

Some sites are of even more recent origin. Metals were never present in especially elevated concentrations at many sites now contaminated by industrial activity, so the evolution of tolerant populations must have occurred subsequent to industrial activity. A zinc smelter in the city of Palmerton, Pennsylvania, for example, first went into operation in 1898, and historical records indicate that severe environmental degradation was evident during the first two decades of the 20th century (Jordan 1975). Zinc ores were brought to Palmerton from New Jersey and the smelter area does not have any naturally occurring enrichment. This site harbors both metal tolerant races of two moss species (*Ceratodon purpureus*: Jules and Shaw 1994; *Funaria hygrometrica*: Shaw 1990), and numerous populations of another moss that is endemic to metal-enriched sites (Shaw and Beer 1989). The latter is widespread in the northern hemisphere and almost certainly did not evolve in situ on contaminated soils near Palmerton, but the ecotypes of common species must have evolved during this century.

Copper mine tailings at Copperpolis, California, harbor a widespread species of *Mimulus* found in the western United States, *M. guttatus*. An endemic, *M. cupriphilus*, is known from only two mines in the immediate vicinity of Copperopolis (Macnair et al. 1989). Macnair et al. (1989) interpret *M. cupriphilus* as a recent derivative of *M. guttatus*. The endemic differs morphologically from *M. guttatus* in having more highly branched stems and smaller flowers. Ecologically, *M. cupriphilus* flowers earlier under uniform conditions (and in the field), apparently exhibits higher selfing rates, and devotes more of its resources to flower and seed production. Genetic analyses indicate that the two taxa are distinguished by at least four polygenic and two major gene systems. Epistatic interactions appear to be involved in some of the character differences between *M. guttatus* and *M. cupriphilus*, showing that the species are now characterized by different patterns of genic interaction. Both are copper tolerant and crossing studies indicate that the same genes are probably responsible for tolerance in the endemic and its putative ancestor (Macnair and Cumbes 1989). The mines where *M. cupriphilus* is endemic were worked as recently as the 1950s. Macnair and Cumbes (1989) suggest that the species probably originated in the last 150 years, and quite possibly during the last 50 years.

Wu et al. (1975) investigated copper tolerance in populations of *Agrostis stolonifera* growing at a series of contaminated sites near Prescot, England, that have been subject to aerial deposition of copper since about 1900. Plants had occupied the sites for variable lengths of time ranging from 5 to 70 years. Populations of *A. stolonifera*, even in sites that were only five years old, exhibit elevated mean levels of genetically-based copper tolerance. The frequencies of tolerant individuals increased with age of the site. Very rapid evolution of tolerant populations has also been documented under electricity pylons in England (Al-Hiyaly et al. 1990). Some of these populations are estimated to be no more than 30 years old (Al-Hiyaly et al. 1993).

The fact that the evolution of tolerance can occur very rapidly is not surprising in light of the evidence that tolerant plants can be selected from normal populations in a single generation (e.g., Gartside and McNeilly 1974). Indeed, colonization of severely contaminated habitats appears in most cases to require tolerance, so at least the initial evolution of a tolerant race is the only "option" for a species to exist in such habitats. Factors that lead to even more substantial differentiation of mine populations, such as the origin of a new species with different morphology and ecological patterns (as in *M. cupriphilus*), are less clear and clearly less common. Nevertheless, over geological time spans, speciation may be a "frequent" long term consequence of evolution in metal-enriched habitats.

## Evolutionary Processes in Relation to Sources of Contamination

Population-level processes involved in the evolution of tolerant plants depend on the nature and origin of contamination. Atmospheric deposition of metals from industrial activity often contaminates intact communities that were not previously subject to enrichment. Initial stages of tolerance evolution and community change involve significant and changing biotic interactions as well as the direct selective advantage of tolerant plants in relation to an increasingly toxic substrate. Selection on adult plants for survival is primary, although long-term persistence of tolerant populations must also involve selection on patterns of reproduction including flower and seed (or spore) formation, seedling (or sporeling) survival, and growth to maturity in subsequent generations. Selection is thus a multistage process and not a simple elimination of nontolerant plants, although differential survival of mature plants is the first stage.

Additional variables that will affect processes of evolution in populations subject to atmospheric deposition include the physical extent and level of enrichment. Contamination may be of regional scope at relatively low levels or of more limited extent with more concentrated deposition. Such differences can be expected to result in different patterns of tolerance evolution caused by different patterns of selection on the plants. It is also possible that low or moderate levels of tolerance, sufficient for survival at sites with low levels of contamination, are effected by different mechanisms and genes compared to the high levels of tolerance needed for survival and reproduction on highly contaminated soils.

The processes of colonization and establishment are primary in the evolution of tolerant populations on extremely contaminated mine tailings. Tailings initially have no plants growing on them and biotic interactions among early colonizers are minimal. Successful seed (or spore) germination and seedling (or sporeling) growth are prerequisite to persistence. Selection on competitive abilities is likely to be at a minimum, at least during initial stages of the evolutionary process. Moreover, the spatial limits of metal-enrichment are typically sharp, and relatively small islands of contamination are surrounded by more or less normal vegetation. Gene flow into metal tolerant populations from surrounding nontolerant conspecifics is an important component of tolerance evolution (McNeilly 1968; see also Postma and Groenendijk, chapter 5; Weis et al. chapter 3).

Selection for tolerance is extreme during colonization of contaminated soils (McNeilly and Bradshaw 1968), and virtually all plants that persist show elevated levels of tolerance. Nevertheless, selection would continue to operate on tolerance levels in subsequent generations. Gene flow via pollen from surrounding uncontaminated sites tends to "dilute" tolerance levels of populations on mine tailings (McNeilly and Bradshaw 1968). Furthermore, the effects of gene flow are likely to change over time (Antonovics 1968). The first plants to colonize tailings occur in low densities and are more likely to receive pollen from nontolerants off the mine than are later colonists that are surrounded by other tolerant individuals.

Patterns of selection on plants growing on contaminated mine tailings are complex. Progeny grown from seeds collected from plants of *A. capillaris* growing on copper-enriched tailings at Drws y Coed in Wales were more variable in tolerance than the adult population from which they were sampled (McNeilly and Bradshaw 1968). Progeny less tolerant than the parents may result from gene flow from off the mine, but the occurrence of progeny with higher levels of tolerance than present in the parental population suggests that the most highly tolerant plants may be selected against. Selection on tolerance in mine habitats, therefore, appears to be stabilizing (McNeilly and Bradshaw 1968).

Selection on other characteristics related to survival and reproduction in mine habitats may show patterns that differ from those characterizing tolerance itself (Antonovics and Bradshaw 1970). Independent selection pressures acting on different characters are likely to cause high levels of mortality among mine plants (Antonovics 1972; 1968).

Adjacent to the mine, nontolerant plants receive gene flow from tolerant mine plants. McNeilly and Bradshaw (1968) found that seed progeny from such plants exhibit tolerance levels higher than the parental population of established adults, suggesting that tolerant individuals are selected against in uncontaminated habitats. With regard to tolerance per se, selection in populations of *A. capillaris* adjacent to the mine appears to be directional (i.e., toward decreased tolerance). In contrast to the situation on the tailings where selection for tolerance is direct, selection against tolerant plants on adjacent normal soil is indirect and appears to result from interactions between tolerant and nontolerant plants (McNeilly 1968).

### Natural Selection and Plant Life Histories

Selection at the seed and seedling stage is crucial and is likely to be intense (McNeilly) 1968; Walley et al. 1974; Wu 1990). McNeilly (1968) found that when seeds are germinated on contaminated soils, nontolerant plants die as soon as the emerging radicle contacts toxic soil. Root growth in both tolerant and nontolerant plants is strongly affected by substrate metals and has often been used experimentally to measure the effects of metals on plants. Root growth is generally presumed to be correlated with overall vegetative viability, and, by extension, plant fitness. Walley et al. (1974) found a close correlation between root growth and plant height for plants grown from seeds on toxic soil. The relationship between root growth and long-term plant fitness is complex, however, and few studies have explicitly explored the situation. Experiments by Duek et al. (1987), for example, demonstrated complex interactions between applied copper or zinc and overall vegetative growth. Root growth per se was not measured, but it did not appear that reproductive success could be predicted from any one measure of vegetative performance.

At the site of a zinc smelter in Palmerton, Pennsylvania, most woody angiosperms have been eliminated from toxic areas surrounding the smelters (Fig. 1). A few species, however, persist in a vegetative state with little or no reproduction. *Smilax* sp., for

**FIG. 1.** *Smilax* sp. on metal-contaminated soil near Palmerton, Pennsylvania. The plants produce extensive vegetative growth each year but die back, and rarely, if ever, form flowers.

example, typically produces considerable vegetative growth, dies back, and then continues each year with a new burst of vegetative growth (Fig. 1). Such plants do not appear to form flowers (or seeds), and the relationship between differing degrees of vegetative growth (among conspecific plants in the area), including root growth, and long term fitness is not clear.

Fitness is a multiplicative function of sequential life history stages. Components of vegetative growth include seed germination, seedling survival, and survival to reproductive maturity. Life history components that determine reproductive success include flower production, pollen and ovule production, fertilization rates, and seed maturation. Most of these traits relate to the performance of the diploid sporophytic phase of the life cycle, but selection could also occur among gametophytes. In angiosperms, gametophytic selection can occur among pollen tubes competing for a limited number of ovules, and among ovules competing for resources during fruit maturation.

Plants growing in metal-contaminated environments may contain elevated metal concentrations in their flowers, including the pistil, and selection may occur among gametophytes as they grow through the style. Because pollen are haploid and population sizes (relative to the sporophyte generation) are huge, selection among gametophytes for metal tolerance could contribute to the rapid evolution of tolerance (Searcy and Mulcahy 1985a). For gametophytic selection to effect an increase in tolerance among diploid sporophytes in a population, there must be a genetic correlation between the expression of tolerance in haploid gametophytes and their diploid sporophytic off-spring. Data from a wide range of studies suggest that in angiosperms, the majority of genes expressed in diploids are also expressed in pollen (Ottaviano and Mulcahy 1989). Correlated tolerances (in gametophytes and sporophytes) have been demonstrated for herbicide resistance, heavy metals, and fungal toxins (Bino et al. 1987; Sari Gorla et al. 1989; Smith and Moser 1985).

Searcy and Mulcahy (1985a) found that increasing concentrations of copper in culture medium reduced pollen germination from both tolerant and nontolerant plants of *M. guttatus*, but germination of pollen from tolerant plants was reduced significantly less. Pollen from three families collected from mine wastes in Copperpolis, California, had germination maxima on media with added copper. A comparison of in vitro germination and growth of pollen from zinc tolerant *S. dioica* and nontolerant *S. alba* showed similar patterns. Germination of pollen from tolerant plants was enhanced by added zinc, and such pollen showed higher tolerance to increasing concentration in terms of both germination and growth.

Direct observations of in vitro pollen tube growth suggest that microgametophyte competition is unlikely, but first estimates were based on comparing growth rates in different flowers (Searcy and Mulcahy 1985b). Nevertheless, seed production declined with increasing copper concentration in flowers when pollen was derived from a sensitive plant but not when derived from a tolerant plant.

Searcy and Macnair (1990) tested whether lower seed production results from reduced fertilization by sensitive pollen or from abortion during seed maturation. Isogenic lines of *M. guttatus* differing in tolerance were used as pollen donors and recipients. When pollen came from a sensitive source, the number of seeds per capsule was reduced on average by 22% in plants supplied with 2 ppm copper and 26% in plants supplied with 4 ppm. Recipients differed in the average number of ovules they produced per capsule, so Searcy and Macnair (1990) used the ratio of the number of mature seeds/total ovules to compare patterns of seed output in different plants and treatments. The number of unfertilized ovules was also estimated. Thus, they were able to distinguish reductions in seed number that resulted from reduced fertilization, as opposed to increased seed abortion (mature seeds/zygotes).

Flower copper concentration accounted for 32% of the variance in seed/ovule ratio when pollen came from a sensitive source, but for only about 1% when pollen came from tolerant sources. There were no differences in percent fertilization by tolerant versus nontolerant donors in relation to metal treatment. There were, however, substantial effects of pollen donor (tolerant, nontolerant) and treatment (copper, control) on the ratio of mature seeds/zygotes. Pollen from tolerant plants yielded a lower seed/zygote ratio in the control treatment than did pollen from nontolerants, but the opposite was true in the copper treatments. Reductions relative to the control for sensitive pollen in pollen-enriched pistils was 14% in the 2 ppm copper treatment and 24% in the 4 ppm treatment. It appeared that seed abortion occurred during several developmental stages, but there were no differences in abortion between maternal plants homozygous versus heterozygous for the major tolerance gene.

Searcy and Macnair (1993) expanded the previous work by applying pollen from tolerant and nontolerant plants in mixture to the same flowers. "Heavy" and "light" pollinations were conducted to determine if microgametophyte competition, conditioned by the relative degree of copper tolerance of donor pollen, occurs within the pistil. Searcy and Macnair (1993) reasoned that if microgametophyte competition occurs in the pistil, they would observe a difference in response to the treatments (copper versus control) for the two pollination intensities. Under competitive conditions (heavy pollination intensities), the more rapidly growing, presumably tolerant, pollen would fertilize more ovules as flower copper concentration increases. Under light pollen intensities, both types of pollen would fertilize ovules because in the absence of pollen competition, the ability to effect fertilization is not influenced by copper in the pistil (Searcy and Macnair 1990). Thus, light pollinations would yield a larger decrease in at least some measures of reproductive success than heavy pollinations.

Searcy and Macnair (1993) were not able to detect any evidence of pollen competition (no significant pollination intensity by treatment interaction). Selection appears to occur in the pistil subsequent to pollination, as suggested by the previous work of Searcy and Macnair (1990). In 10 out of 13 crosses performed, there was a higher proportion of tolerant progeny from maternal plants supplied with copper than from control plants. Overall, there was about a 7% increase in copper tolerant progeny when pollen recipients were grown in 4 ppm copper. All progeny were grown on normal soil for six to eight weeks prior to tolerance testing to eliminate nongenetic effects on tolerance induced by the maternal environment in which the seeds developed. Thus, there appears to be a genetic response to pre-emergent selection on developing seeds in copper-enriched flowers.

In bryophytes, initial life-history stages analogous to seeds and seedlings are the spores and sporelings. In contrast to the seed plants, however, these bryophyte stages are haploid. The immature stage of moss gametophytes (the free-living stage of the life cycle) consists of filamentous protonemata on which leafy stems are formed. In order for plants to form reproductive structures (archegonia and antheridia [gametangia]), and, following fertilization, sporophytes with spores, the spores must germinate, protonemata must grow and form stems, and the stems must form gametangia. Initial selection is obviously on spore germination but success of plants colonizing contaminated habitats depends also on subsequent life-history stages. Spore germination is less sensitive to substrate metals than is growth of the protonemata (Shaw et al. 1987). Stem formation is not well correlated with protonemal growth, and many plants that are capable of forming relatively well-developed protonemata do not form mature stems (Shaw 1990, 1991). Even metal tolerant plants of the moss, *F. hygrometrica*, reach a threshold of contamination in which plant stature and vegetative growth rates are severely reduced (Shaw and Bartow 1992).

In contrast to the situation in flowering plants, selection for metal tolerance in bryophytes will occur primarily in the haploid gametophyte generation because it is the gametophytes that are free-living. Mosses growing in areas subject to metal deposition often do not form sporophytes or form them at reduced frequencies

(Longton 1985; Shaw 1990). Such reductions could result from reduced gametangial formation, reduced fertilization, and/or sporophyte abortion. Metal tolerant plants of *C. purpureus* did not form gametangia when grown on metal-enriched soil, but did so on soil from the smelter site where they originated (Jules and Shaw 1994). Conspecific plants from normal sites, in contrast, formed gametangia on control but not smelter soil. Data presented by Longton (1985) suggest that sporophyte abortion may be increased in moss populations subject to aerial deposition of metals. *Bryum bicolor* does not form sporophytes on metal-contaminated soil, but plants grown experimentally on soil enriched with metals form more abundant asexual propagules than those grown on control soil (Shaw 1992).

## GENETIC LIMITATIONS ON TOLERANCE EVOLUTION

It is a fundamental premise of all natural selection theories that in the absence of genetic variation for a trait, there can be no response to selection (Fisher 1930). Moreover, it has long been recognized that some plant species evolve tolerance to metals and others do not (Antonovics et al. 1971). In general, for species that occur on metal-contaminated mine tailings, at least a low frequency of tolerant individuals can be detected in normal populations (Bradshaw 1984, 1991; Macnair 1987). Many families of angiosperms include species that evolve metal tolerance, and it does not appear that the propensity to occur on contaminated soils, presumably reflecting genetic variance for tolerance, is taxonomically biased (Antonovics et al. 1971). Likewise, metal tolerant bryophytes occur in many families of mosses and liverworts (Shaw 1990). Genetic limitations on the colonization of mine tailings (and other types of "industrial soil") must also involve the occurrence of genetic variation for other severe environmental features of such habitats. Nevertheless, the occurrence of genes for tolerance must be the ultimate limiting factor.

Gartside and McNeilly (1974) found that normal populations of *A. capillaris* (= *A. tenuis*) and *Dactylis glomerata*, two species commonly found on copper-contaminated soils, contained about 0.08% tolerant individuals, whereas no fully tolerant plants (with levels comparable to mine site plants) were detected in normal populations of *Poa trivialis*, *Lolium perenne*, *Arrhenatherum elatius*, and *Cynosurus cristatus*. The latter do not typically occur on copper contaminated soils, and the populations tested apparently lack the genetic variation to evolve full tolerance. A low frequency of plants did express partial tolerance (7–20% of fully tolerant conspecifics).

Al-Hiyaly et al. (1993) tested the hypothesis that genetic variation for tolerance is the major limitation on tolerance evolution, even on a very local scale. They compared the occurrence and levels of zinc tolerance in plants of *A. capillaris* under and adjacent to five electricity pylons (subject to zinc leachate). Patterns of tolerance in plants under each pylon were related to the availability of genetic variation in adjacent plants. One pylon, for example, did not have tolerant *A. capillaris*, and no tolerant individuals could be detected in an immediately adjacent population on normal soil.

Since selection, at least for metal tolerance per se, was presumed to be uniform among the pylons, differences between them in the occurrence of tolerant plants must have been due to stochastic variation in the spatial distribution of tolerance genes in *A. capillaris* (Al-Hiyaly et al. 1993). Although the assumption that selective pressures were comparable among pylons seems reasonable, unfortunately this was not explicitly tested.

## COSTS OF TOLERANCE EVOLUTION

Much indirect evidence suggests that costs are associated with tolerance evolution, i.e., that tolerant plants are at a selective disadvantage relative to "normal" conspecific plants on uncontaminated soil (Fig. 2). Direct tests to elucidate the nature of such costs, however, have been inconclusive.

Two types of observations strongly suggest that tolerance has its costs. One is the frequent observation that although many species include genetically distinct tolerant races on mine tailings, the frequency of tolerant individuals in normal populations is almost always low (<1%; see above, and Macnair 1987). The other is that relatively steep clines for tolerance occur at the boundaries of contaminated soil. If tolerance genes are neutral in uncontaminated sites, they should be more ubiquitous throughout a species range and should diffuse from locally contaminated sites into adjacent areas. Such is not the case in most species.

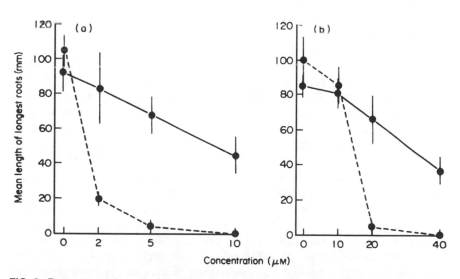

**FIG. 2.** Response of *Agrostis stolonifera* to copper (*a*) and zinc (*b*). The pattern in which tolerant plants (solid line) form less root growth than nontolerants (dashed line) when grown without added metals is commonly observed, and suggests a "cost" associated with tolerance. Redrawn from Wu and Antonovics (1975).

A few species appear to have inherent tolerance of one or more metals and colonize contaminated sites without the evolution of specialized races. Such cases are relatively few, however, and clearly constitute exceptions to the general pattern. Nevertheless, as exceptions, they take on particular interest. One case is *Typha latifolia*, which occurs in a variety of wet, contaminated places. Yet plants growing under conditions of metal-enrichment are no more tolerant than other individuals of the species collected from uncontaminated sites (McNaughton et al. 1974; Taylor and Crowder 1984). Plants of *Andropogon virginicus* growing on lead/zinc mine wastes in Oklahoma exhibit no greater tolerance of mine soil than those from adjacent fields or uncontaminated control sites (Gibson and Risser 1982).

Wu and Antonovics (1976) found that roadside plants of *Plantago lanceolata* in Durham, North Carolina, were more tolerant of lead than nearby plants away from the roadside, but that plants of *Cynodon dactylon* along the same roadside were not genetically differentiated for lead tolerance. Their approach utilized both cloned plants that had been grown before the experiment on uncontaminated soil and seeds collected from field-grown plants. They found that variations in tolerance level were unlikely to be the result of acclimation to the contaminated roadside soils. *Cynodon* appeared to have a higher level of inherent tolerance than *Plantago*, and this allowed it to grow in moderately contaminated sites. The level of tolerance in *C. dactylon* may not be sufficient for it to colonize sites of severe contamination.

Metal tolerance without the evolution of specialized races has also been found in the moss *Bryum argenteum* (Shaw and Albright 1990). Populations of *B. argenteum* from contaminated and uncontaminated sites appear to possess inherently high tolerance of metals (Fig. 3). Typical populations from sites not enriched with metals consist of plants that are more tolerant than unspecialized individuals of *F. hygrometrica*, but less tolerant than genetically specialized races of the latter found growing on contaminated mine tailings (Fig. 3). *B. argenteum* does colonize severely contaminated mine tailings, but the plants are generally weak there and do not form sporophytes (unlike specialized races of *F. hygrometrica*, which are fertile on mine tailings).

Watkins and Macnair (1991) showed that variation in arsenic tolerance is heritable in *A. capillaris* and appears to be controlled by a single major gene, probably with at least one additional interacting gene that affects levels of tolerance. In *H. lanatus*, tolerance is not restricted to sites enriched with arsenic (Macnair et al. 1992). Meharg et al. (1993) surveyed 39 British populations of *H. lanatus* from uncontaminated sites and 11 from sites with arsenic enrichment. They found that over 90% of the plants on arsenic contaminated soils were tolerant, but the mean percentage of tolerant plants in populations not subject to arsenic enrichment was 45.3% (range 15–70%). Meharg and Macnair (1992) showed that reduced uptake of arsenic by tolerant plants of *H. lanatus* is related to a modification of the phosphate uptake system (arsenate acts as a phosphate analog). Although such a modification might seem to place tolerant plants at a disadvantage in nutrient poor (but not arsenic-enriched) habitats, phosphate acquisition by plants may be limited by rates of diffusion into roots, rather than by the rate of phosphate movement across the plasmalemma. Thus, the modified uptake

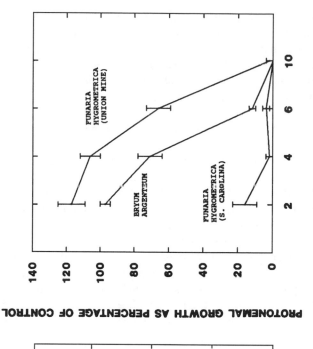

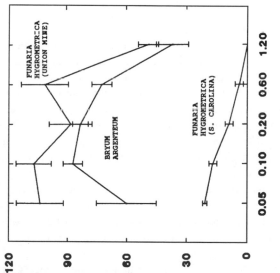

FIG. 3. Response of *Bryum argenteum* to cadmium (left) and copper (right), compared to tolerant and nontolerant races of *Funaria hygrometrica*. Data for *B. argenteum* are based on eight populations that do not differ in tolerance. Redrawn from Shaw (1990).

system involved in arsenic tolerance in *H. lanatus* may not involve a severe cost in terms of mineral nutrition by normal plants. The fact that the gene(s) for arsenic tolerance is relatively common and ubiquitous in *H. lanatus*, including nutrient poor sites, supports that interpretation (Meharg and Macnair 1992).

Sharp clines for tolerance, and sometimes also morphological characters, are often observed at ecotones separating mine tailings and adjacent normal soils (McNeilly and Bradshaw 1968; Antonovics et al. 1971). The position, spatial scale, and shape of clinal variation result from interactions between gene flow and selection on either side of the ecotone. The abruptness of such clines despite gene flow shows that selection for tolerance is strong within the contaminated areas and also suggests that there is significant selection against tolerant individuals on normal soil.

McNeilly and Bradshaw (1968) showed that adult plants of *A. capillaris* growing on Drws y Coed, a copper mine in Wales, had a higher mean tolerance of copper than seed samples collected from the mine population, suggesting that gene flow from nearby nontolerant plants reduced tolerance levels. Strong selection for tolerance must counteract such gene flow. In adjacent pasture populations, the opposite pattern was found; seed samples included about 3% tolerant individuals, yet the frequency of tolerant adults in the population was less than 1%. Selection against tolerants off the mine is the most viable explanation for this pattern. The fact that seed samples from the mine population contained individuals with both higher and lower levels of tolerance compared to the adult population suggests stabilizing selection on the contaminated soils. The near absence of tolerant individuals in adjacent pasture populations suggests that selection is directional (against tolerants) on normal soil.

Modeling the interaction between gene flow and selection in producing clinal variation at mine/pasture ecotones, Antonovics (1968) found that strong selection will maintain population differences on either side, but that the degree of mortality caused by the movement of disadvantageous genes across the ecotone is highest when the favored genes are at relatively low frequency and are not, or incompletely, dominant. Tolerance appears to be dominant in most cases in which the genetics have been adequately investigated. The frequency of the tolerance gene within an area in which pollen movement between plants is likely will be relatively low in early stages of colonization, but would increase rapidly as mine tailings are colonized by tolerant individuals. Off the mine, the introduction of genes for tolerance would cause substantial mortality if selection against tolerant individuals is strong. However, if the cost of tolerance is not associated with tolerance genes, but rather with adaptations for other stress-inducing features of mine habitats, a clear prediction cannot be made without genetic analyses.

An early study by McNeilly (1968) illustrated the importance of gene flow in defining patterns of clinal variation at mine ecotones. At a small lead/zinc mine in Wales, the prevailing wind patterns are oriented longitudinally down a U-shaped glacial valley. The cline for metal tolerance was abrupt at right angles to prevailing wind direction, but was more gradual in the down-wind direction. This study showed that gene flow can substantially affect the shape of the cline. Moreover, the steep cline from the mine tailings into the pasture across the direction of the prevailing wind patterns suggests that selection against tolerant individuals growing on normal soil must be significant.

Antonovics and Bradshaw (1970) showed that clinal patterns across the same ecotone in *Anthoxanthum odoratum* differed among characters. Some characters, such as metal tolerance, flowering time, and the degree of self-fertility, showed abrupt changes. Others, including flag leaf length and culm length, showed more gradual clines, whereas still others, such as flag leaf width and tiller number, showed no clinal variation. Even characters that did not exhibit clinal variation were variable among plants within the mine or pasture, suggesting that the absence of clinal patterns in these features did not result from a lack of genetic variation. Antonovics and Bradshaw (1970) interpreted the patterns they observed as reflecting independent selection pressures on different characters. The fact that most characters were not correlated within sites (mine versus pasture) means that correlated patterns of variation between sites must have resulted from independent selection. Antonovics and Bradshaw (1970) pointed out that independent selection would greatly increase selective mortality (i.e., the overall intensity of selection).

Clinal patterns at mine boundaries strongly suggest that plants growing on metal-contaminated mine tailings are at a disadvantage in normal habitats. Nevertheless, it is not clear that tolerance is the trait, or the only trait, that carries a cost. If selection on a mine favors not only metal tolerant plants, but also plants that are slow growing and small in stature (perhaps because of water and nutrient stress), linkage disequilibrium among genes controlling these traits may develop within the mine population. Plants that disperse from the mine onto adjacent normal soil are characterized by the suite of correlated traits found in mine plants, and any selective disadvantage they encounter may or may not result from the tolerance gene(s) per se.

Macnair and Watkins (1983) used isogenic lines of *M. guttatus* differing only in tolerance to investigate costs associated specifically with the tolerance gene when plants were grown on normal soil. They measured eight characters related to plant viability and reproductive vitality (including competitive ability) and found no evidence of any differences between tolerants and nontolerants. Although these results are interesting, the experiments were undertaken in potted plants grown in a greenhouse and it is possible that any fitness differences that might exist between tolerants and nontolerants could only be detected under different environmental conditions. Field experiments would provide stronger evidence for a lack of cost.

In a detailed study of the distribution of tolerance on and around an abandoned copper mine in Copperpolis, California, Macnair et al. (1993) found that 100% of the plants on the tailings were tolerant. South of the tailings, along a small stream where *M. guttatus* grows, many of the plants were also tolerant. Upstream from the mine, the proportion of tolerant plants was lower, but tolerant individuals segregated among the offspring of naturally pollinated adults. Thus, the study demonstrated strong selection for tolerance on contaminated tailings, but the very gradual clines away from the mine, especially in the direction of presumed gene flow, suggest that selection against tolerants on normal soil is not strong. Steeper clines were found for the degree (as opposed to the presence) of tolerance. Because the degree of gene flow for the presence of tolerance and the degree of tolerance must be equivalent, steeper clines for the latter suggest stronger selection on the latter. If the presence and degree of tolerance are under different genetic control (see above), these observations suggest that selection pressures on the major gene (controlling the presence of tolerance) differs from the pressures on modifiers that underlie variation in the degree of tolerance.

Any costs associated with tolerance could be manifest in altered physiological requirements of tolerant plants and/or reduced competitive abilities of tolerant plants growing in uncontaminated habitats having higher plant densities. Tolerant plants of a number of species tend to be smaller and weaker than nontolerant conspecifics (Antonovics et al. 1971). Lead mine plants of *A. capillaris* exhibit a higher tolerance of low nutrients than plants from uncontaminated sites (Jowett 1959), and this may imply a cost when plants are growing in normal soils without nutrient deficiency. Zinc tolerant plants of *Armeria maritima* and *Anthoxanthum odoratum* appear to require elevated levels of zinc for optimal growth (Antonovics et al. 1971), and it is possible that tolerant plants of some species suffer nutrient deficiency in the absence of elevated micronutrients. Elevated metal concentrations are also required for optimal growth in metal tolerant plants of several mosses (Shaw 1990). This phenomenon could result from tolerance mechanisms that involve sequestration of metals in cell walls (or elsewhere); efficient mechanisms for reducing intracellular uptake could lead to nutrient deficiency on substrates lacking metal enrichment.

McNeilly (1968) compared growth of copper tolerant *A. capillaris*, collected as seed from mine tailings at Drws y Coed in Wales, with nontolerant plants that orginated adjacent to the mine. There was no detectable difference in growth when the plants were grown individually, but when planted in 50:50 mixtures, tolerants formed significantly fewer tillers and formed less total biomass than nontolerants. Wilson (1988) found a correlation between tolerance levels and vegetative growth rates in a sample of eight British populations (four from mines) and suggested that reduced growth rate is a cost associated with metal tolerance. However, it is not clear whether reduced growth rates are selected independently of tolerance in mine populations, nor how growth rate is related to fitness in normal populations.

Cook et al. (1972) showed that in a greenhouse experiment, tolerant plants of *A. odoratum* and *A. capillaris* are competitively inferior to nontolerant conspecifics. Expanding that line of inquiry as a field experiment, Hickey and McNeilly (1975) showed that tolerant individuals of *A. capillaris*, *A. odoratum*, *P. lanceolata*, and *Rumex acetosa*, are weaker competitors than nontolerants of the same species. Relative fitness of tolerants on normal soil ranged from 0.32 in *A. capillaris* to 0.001 for *A. ordoratum*. Antonovics and Bradshaw (1970) showed that the cline for zinc tolerance at the Trelogan Mine in Wales is steeper for *Anthoxanthum* than for *A. capillaris*, as predicted if the costs associated with tolerance are greater for the former than for the latter.

## CONCLUSIONS

The occurrence of plants in metal contaminated habitats continues to provide a natural experiment for investigating evolutionary processes in plant populations. Continuing studies on the genetic basis of metal tolerance are especially needed. Work by Macnair et al. on *M. guttatus* and *H. lanatus* and by Schat et al. on *S. vulgaris*, have substantially clarified the evolution of tolerance by providing insights into genetics underlying the phenomenon. Similar work on other species is required to determine if the genetic basis of tolerance is comparable in other taxa and for other metals (see Postma and Groenendijk, chapter 5). More extensive sampling of populations is needed to assess whether additional tolerance genes can be detected within particular species. The application of molecular mapping techniques to assess the number and location of loci involved in tolerance evolution constitutes a promising avenue of research.

The issue of costs associated with tolerance needs focused research. Indirect evidence from the distribution of tolerance in uncontaminated habitats and from clinal variation at ecotonal areas separating contaminated from normal sites suggests that costs associated with the "tolerance syndrome" are substantial. Direct experimental investigations however, are few, and inconclusive. In addition, future research must separate costs associated with other traits characterizing mine plants, with tolerance per se. The availability of isogenic lines differing only in tolerance, as developed and employed by Macnair et al., permits unprecedented investigations on pleiotropic effects of the tolerance genes. The negative results of Macnair and Watkins (1983), in which no deleterious characteristics were found to be associated with the tolerance genes, might suggest that it is adaptation to the suite of other hostile mine site environmental factors that carries a significant cost. Nevertheless, the near absence of metal tolerance from normal populations of most species does indicate that any costs must involve pleiotropic effects of tolerance to genes that are tightly linked with tolerance or to specific gene combinations that are selected in severe habitats that are also contaminated with metals.

Recent work on fitness costs associated with the evolution of antibiotic resistance in bacteria suggests that such costs can be reduced or even eliminated by natural selection in organisms living in antibiotic-free habitats (Lenski 1997). Moreover, the magnitude and nature of costs associated with antibiotic resistance seem to be dependent on the genetic background in which the resistance gene occurs. In bacteria, coevolution between resistance genes and their genetic background can occur within a few hundred generations, which takes only a month or two. Although comparable numbers of generations would take a century or more in most plants, it is possible that selection could ultimately reduce any initial costs that are associated with tolerance. This possibility is worthy of study.

Wu and Antonovics (1976) quoted from a 1972 U.S. National Academy of Sciences Report questioning the deleterious impact of lead contamination on plants, given that plants evolve tolerance to environmental contamination by lead. They point out that it is precisely the deleterious effects of lead that result in selection for tolerance in natural populations. In addition, insidious costs associated with tolerance could limit the ability of species to adapt to future, unanticipated environmental changes, and their importance should not be underestimated.

## ACKNOWLEDGMENTS

Support for work on metal tolerances in bryophytes was funded by NSF grant DEB-9407937. I am grateful to J. Antonovics for discussions during preparation of this manuscript and to M. Macnair for providing unpublished information about ongoing genetic analyses.

## REFERENCES

Al-Hiyaly, S. A. K., T. McNeilly, and A. D. Bradshaw. 1990. The effect of zinc contamination from electricity pylons. Contrasting patterns of evolution in five grass species. *New Phytologist* 114:183–190.

Al-Hiyaly, S. A. K., T. McNeilly, A. D. Bradshaw, and A. M. Mortimer. 1993. The effect of zinc contamination from electricity pylons. Genetic constraints on selection for zinc tolerance. *Heredity* 70:22–32.

Aniol, A., and G. P. Gustafson. 1990. Genetics of tolerance in agronomic plants. In *Heavy metal tolerance in plants: Evolutionary aspects*, ed. A. J. Shaw, 255–267. Boca Raton: CRC Press.

Antonovics, J. 1968. Evolution in closely adjacent plant populations VI. Manifold effects of gene flow. *Heredity* 23:507–524.

Antonovics, J. 1972. Population dynamics of the grass *Anthoxanthum odoratum* on a zinc mine. *Journal of Ecology* 60:351–365.

Antonovics, J. 1975. Metal tolerance in plants: Perfecting an evolutionary paradigm. *International Conference on Heavy Metals in the Environment* 169:186.

Antonovics, J., and A. D. Bradshaw. 1970. Evolution in closely adjacent plant populations VIII. Clinal patterns at a mine boundary. *Heredity* 25:349–362.

Antonovics, J., A. D. Bradshaw, and R. G. Turner. 1971. Heavy metal tolerance in plants. *Advances in Ecological Research* 7:1–85.

Baker, A. J. M. 1987. Metal tolerance. *New Phytologist* 106(Suppl):93–111.

Bino, R. J., J. Hille, and J. Franken. 1987. Kanamycin resistance during in vitro development of pollen from transgenic plants. *Plant Cell Reports* 6:333–336.

Bradshaw, A. D. 1984. The importance of evolutionary ideas in ecology—and vice versa. In *Evolutionary Ecology*, ed. B. Shorrocks, 1–26. Oxford: Blackwell.

Bradshaw, A. D. 1991. Genostasis and the limits to evolution. *Philosophical Transactions of the Royal Society of London Series B* 333:289–305.

Bradshaw, A. D., and T. McNeilly. 1981. *Evolution and pollution*. London: Edward Arnold.

Brooks, R. R., and F. Malaisse. 1985. *The heavy metal-tolerant flora of southcentral Africa*. Rotterdam: AA Balkema.

Cook, S. C. A., C. Lefèbvre, and T. McNeilly. 1972. Competition between metal tolerant and normal plant populations on normal soil. *Evolution* 26:366–372.

Duek, T. A., H. G. Wolting, D. R. Moet, and F. J. M. Pasman. 1987. Growth and reproduction in *Silene cucubalus* Wib. intermittently exposed to low concentrations of air pollutants, zinc and copper. *New Phytologist* 105:633–645.

Farago, M. E. 1981. Metal tolerant plants. *Coordinated Chemical Review* 36:155–182.

Fisher, R. A. 1930. *The genetical theory of natural selection*. Oxford: Clarendon Press.

Gartside, D. W., and T. McNeilly. 1974. The potential for the evolution of heavy metal tolerance in plants II. Copper tolerance in normal populations of different plant species. *Heredity* 32:335–348.

Gibson, D. J., and P. G. Risser. 1982. Evidence for the absence of ecotypic development in *Andropodon virginicus* (L.) on metaliferous mine wastes. *New Phytologist* 92:589–599.

Hickey, D. A., and T. McNeilly. 1975. Competition between metal tolerant and normal plant populations: A field experiment on normal soil. *Evolution* 29:458–464.

Humphries, M. O., and M. K. Nicholls. 1984. Relationships between tolerance to metals in *Agrostis capillaris* L. (*A. tenuis* Sibth.). *New Phytologist* 98:177–190.

Jordan, M. L. 1975. Effects of zinc smelter emissions and fire on a chestnut-oak woodland. *Ecology* 56:78–91.

Jowett, D. 1959. Adaptation of a lead-tolerant population of *Agrostis tenuis* to low soil fertility. *Nature* 184:43.

Jules, E. S., and A. J. Shaw. 1994. Adaptation to metal-contaminated soils in populations of the moss, *Ceratodon purpureus*: Vegetative growth and reproductive expression. *American Journal of Botany* 82:8–17.

Lande, R. 1983. The response to selection on major and minor mutations affecting a metrical trait. *Heredity* 50:47–65.

Lenski, R. E. 1997. The cost of antibiotic resistance—from the perspective of a bacterium. In *Antibiotic resistance: Origins, evolution, selection and spread*, ed. D. J. Chadwick and J. Goode, 131–151. Chichester, UK: Wiley. [*Ciba Foundation Symposium Volume* 207]

Longton, R. E. 1985. Reproductive biology and susceptibility to air pollution in *Pleurozium schreberi* (Brit.) Mitt (Musci) with particular reference to Manitoba, Canada. *Monographs in Systematic Botany from the Missouri Botanical Garden* 11:51–69.

Macnair, M. R. 1981. The tolerance of higher plants to toxic materials. In *Genetic consequences of manmade change*, eds. J. A. Bishop and L. M. Cook, 177–208. London: Academic Press.

Macnair, M. R. 1983. The genetic control of copper tolerance in yellow monkey flower, *Mimulus guttatus*. *Heredity* 50:283–293.

Macnair, M. R. 1987. Heavy metal tolerance in plants: A model evolutionary system. *Trends in Ecology and Evolution* 2:354–359.

Macnair, M. R. 1990. The genetics of metal tolerance in natural populations. In *Heavy metal tolerance in plants: Evolutionary aspects*, ed. A. J. Shaw, 235–253. Boca Raton: CRC Press.

Macnair, M. R. 1991. Why the evolution of resistance to anthropogenic toxins normally involves major gene changes: The limits to natural selection. *Genetica* 84:213–219.

Macnair, M. R. 1993. The genetics of metal tolerance in vascular plants. Tansley Review 49. *New Phytologist* 124:541–559.

Macnair, M. R., and Q. J. Cumbes. 1989. The genetic architecture of interspecific variation in *Mimulus*. *Genetics* 122:211–222.

Macnair, M. R., Q. J. Cumbes, and A. A. Meharg. 1992. The genetics of arsenate tolerance in Yorkshire fog, *Holcus lanatus* L. *Heredity* 69:325–335.

Macnair, M. R., V. E. Macnair, and B. E. Martin. 1989. Adaptive speciation in *Mimulus*: An ecological comparison of *M. cupriphilus* with its presumed progenitor, *M. guttatus*. *New Phytologist* 112:269–279.

Macnair, M. R., S. E. Smith, and Q. J. Cumbes. 1993. Heritability and distribution of variation in degree of copper tolerance in *Mimulus guttatus* at Copperopolis, California. *Heredity* 71:445–455.

Macnair, M. R., and A. D. Watkins. 1983. The fitness of the copper tolerance gene of *Mimulus guttatus* in uncontaminated soil. *New Phytologist* 95:133–137.

McNaughton, S. J., T. C. Folsom, T. Lee, F. Park, C. Price, D. Roeder, J. Schmitz, and C. Stockwell. 1974. Heavy metal tolerance in *Typha latifolia* without the evolution of tolerant races. *Ecology* 55:1163–1165.

McNeilly, T. 1968. Evolution in closely adjacent plant populations III. *Agrostis tenuis* on a small copper mine. *Heredity* 23:99–108.

McNeilly, T., and J. Antonovics. 1968. Evolution in closely adjacent plant populations IV. Barriers to gene flow. *Heredity* 23:205–218.

McNeilly, T., and A. D. Bradshaw. 1968. Evolutionary processes in populations of copper tolerant *Agrostis tenuis* Sibth. *Evolution* 22:108–118.

Meharg, A. A., Q. J. Cumbes, and M. R. Macnair. 1993. Pre-adaptation of Yorkshire fog, *Holcus lanatus* L. (Poaceae) to arsenate tolerance. *Evolution* 47: 313–316.

Meharg, A. A., and M. R. Macnair. 1992. Polymorphism and physiology of arsenate tolerance in *Holcus lanatus* L. from an uncontaminated site. *Plant and Soil* 146:219–225.

Ottaviano, E., and D. L. Mulcahy. 1989. Genetics of angiosperm pollen. *Advances in Genetics* 29:1–64.

Nicholls, M. K., and T. McNeilly. 1982. The possible polyphyletic origin of copper tolerance in *Agrostis tenuis* (Gramineae). *Plant Systematics and Evolution* 140:109–117.

Prat, S. 1934. Die Erblichkeit der Resistenz gegen Kupfer. *Berichte der Deutschen Botanischen Gesellschaft* 52:65–67.

Purvis, O. W., and C. Halls. 1996. A review of lichens in metal-enriched environments. *Lichenologist* 28:571–601.

Sari Gorla M. E., E. Ottaviano, E. Frascaroli, and P. Lande. 1989. Herbicide-tolerant corn by pollen selection. *Sexual Plant Reproduction* 2:65–69.

Schat, H., E. Kuiper, W. M. Ten Bookum, and R. Vooijs. 1993. A general model for the genetic control of copper tolerance in *Silene vulgaris*: Evidence from crosses between plants from different environments. *Heredity* 70:142–147.

Schat, H., and W. M. Ten Bookum. 1992. Genetic control of copper tolerance in *Silene vulgaris*. *Heredity* 68:219–229.

Searcy, K. B., and M. R. Macnair. 1990. Differential seed production in *Mimulus guttatus* in response to increasing concentrations of copper in the pistil by pollen from copper tolerant and sensitive sources. *Evolution* 44:1424–1435.

Searcy, K. B., and M. R. Macnair. 1993. Developmental selection in response to environmental conditions of the maternal parent in *Mimulus guttatus*. *Evolution* 47:13–24.

Searcy, K. B., and D. L. Mulcahy 1985a. The parallel expression of metal tolerance in pollen and sporophytes of *Silene dioica* (L.) Clairv., *S. alba* (Mill.) Krause and *Mimulus guttatus* DC. *Theoretical and Applied Genetics* 69:597–602.

Searcy, K. B., and D. L. Mulcahy, 1985b. Pollen tube competition and selection for metal tolerance in *Silene dioica* (L.) Clairv. (Caryophyllaceae) and *Mimulus guttatus* (Scrophulariaceae). *American Journal of Botany* 72:1695–1699.

Shaw, A. J. ed. 1990. *Heavy metal tolerance in plants: Evolutionary aspects*. Boca Raton: CRC Press.

Shaw, A. J. 1991. Ecological genetics, evolutionary constraints, and the systematics of bryophytes. *Advances in Bryology* 4:29–74.

Shaw, A. J. 1992. Genetic and environmental effects on morphology and asexual reproduction in the moss, *Bryum bicolor. Bryologist* 93:1–6.

Shaw, A. J., and D. L. Albright. 1990. Potential for the evolution of heavy metal tolerance in *Bryum argenteum*, a moss II. Generalized tolerances among diverse populations. *Bryologist* 93:187–192.

Shaw, A. J., J. Antonovics, and L. E. Anderson. 1987. Inter- and intraspecific variation of mosses in tolerance to copper and zinc. *Evolution* 41:1312–1325.

Shaw, A. J. and S. M. Bartow. 1992. Genetic structure and phenotypic plasticity in proximate populations of the moss, *Funaria hygrometrica. Systematic Botany* 17:257–271.

Shaw, A. J., and S. C. Beer. 1989. *Scopelophila cataractae* in Pennsylvania. *Bryologist* 92:112–115.

Smith, G. A., and H. S. Moser. 1985. Sporophyte-gametophyte herbicide tolerance in sugarbeet. *Theoretical and Applied Genetics* 71:231–237.

Smith, S. E., and M. R. Macnair. 1998. Hypostatic modifiers cause variation in degree of metal tolerance in *Mimulus guttatus*. *Heredity* 80:760–768.

Taylor, G. J., and A. A. Crowder. 1984. Copper and zinc tolerance in *Typha latifolia* clones from contaminated and uncontaminated environments. *Canadian Journal of Botany* 62:1304–1308.

Turreson, G. 1922. The genotypical response of the plant species to the habitat. *Hereditas* 3:211–350.

Walley, K. A., M. S. I. Khan, and A. D. Bradshaw. 1974. The potential for evolution of heavy metal tolerance in plants I. Copper and zinc tolerance in *Agrostis tenuis*. *Heredity* 32:309–319.

Watkins, A. J., and M. R. Macnair. 1991. Genetics of arsenic tolerance in *Agrostis capillaris* L. *Heredity* 66:47–54.

Wickland, D. E. 1983. Vegetation patterns on derelict heavy metal mine sites in the North Carolina Piedmont. Ph.D. dissertation, Chapel Hill: University of North Carolina.

Wilson, J. B. 1988. The cost of heavy metal tolerance: An example. *Evolution* 42:408–413.

Wu, L. 1990. Colonization and establishment of plants in contaminated sites. In Heavy metal tolerance in plants: Evolutionary aspects, ed. A. J. Shaw, 269–284. Boca Raton: CRC Press.

Wu, L., and J. Antonovics. 1975. Zinc and copper uptake by *Agrostis stolonifera*, tolerant to both zinc and copper. *New Phytologist* 75:231–237.

Wu, L., and J. Antonovics. 1976. Experimental ecological genetics in *Plantago* II. Lead tolerance in *Plantago lanceolata* and *Cynodon dactylon* from a roadside. *Ecology* 57:205–208.

Wu, L., A. D. Bradshaw, and D. A. Thurman. 1975. The potential for the evolution of heavy metal tolerance in plants III. The rapid evolution of copper tolerance in *Agrostis stolonifera*. *Heredity* 34:165–187.

Genetics and Ecotoxicology
Edited by V. E. Forbes
Copyright © 1999 Taylor & Francis

3

# Mercury Tolerance, Population Effects, and Population Genetics in the Mummichog, *Fundulus heteroclitus*

Judith S. Weis, Nicolay Mugue, and Peddrick Weis

**Abstract.** Embryos of *Fundulus heteroclitus* from clean environments demonstrate a wide variation in teratogenic response to methylmercury, with tolerance associated with traits of the female. Reduced uptake of the toxicant through the chorion and faster development time appear to be mechanisms responsible for the enhanced tolerance. Embryos from a contaminated site demonstrate enhanced tolerance to meHg exposure. Tolerance is not seen in larvae after hatching, nor in adults, which exhibit decreased tolerance and stress presumably caused by continual exposure to toxicants. Adults from the polluted population do not grow as well or live as long as the reference population but become reproductive sooner, an evolutionary strategy for perpetuating a population in a stressful location. A possible reason for the reduced growth in the polluted population may be a poor diet resulting from decreased prey capture ability, which appears to be due to neurotoxic chemicals but may also have genetic components. Two *F. heteroclitus* subspecies have been recognized, with the separation point in New Jersey. Extensive allozyme and mitochondrial DNA studies revealed in more detail the intergradation zone of the two subspecies and identified an intermediate form in northern New Jersey, which may have arisen through hybridization of the two races. No genetic markers clearly associated with pollution and enhanced tolerance could be distinguished, however.

**Keywords.** Allozyme; heavy metals; mitochondrial DNA; subspecies; teratogenic effects.

## INTRODUCTION

Tolerance to contaminants has often been observed in chronically exposed populations. (In this paper we are using the terms "tolerance" and "resistance" interchangeably.) There are physiological mechanisms by which individuals can acquire increased tolerance as a result of pre-exposure to a given contaminant. For example, metallothioneins, metal-binding proteins synthesized in response to metal exposure, allow the organism to tolerate higher concentrations of the metal than one not pre-exposed (Pascoe and Beattie 1979). By binding metals, metallothioneins prevent them from binding to various enzymes that are important in biochemical reactions. There have been many studies (for example, Bradley et al. 1985) in which organisms (e.g., fish) have been pre-exposed to a metal and subsequently exhibited enhanced resistance to it as a result of synthesizing more metallothioneins. A feature of such physiologically-based tolerance is that when organisms are placed back in clean water for a period of time, the tolerance is rapidly lost (Benson and Birge 1985). Another mechanism for increasing metal resistance is by sequestration of the metals into storage granules. In addition to these short-term physiological mechanisms, there are also evolutionary, genetic mechanisms by which selection for more tolerant individuals can occur in polluted environments, causing populations residing there to have enhanced tolerance to the contaminants (Klerks and Weis 1987). The contaminant can act as a selective agent, eliminating those in the population that are more susceptible to it. Such selection can lead to a more tolerant population, which would have less genetic variability within it.

## EMBRYOLOGICAL TOLERANCE TO METHYLMERCURY

### High Variability in Clean Environments

Our research initially focused on teratological effects of methylmercury (meHg), a potent teratogenic and neurotoxic chemical, in embryos of the mummichog, *Fundulus heteroclitus*. This fish species is very abundant in the estuaries of the Atlantic coast of North America from Canada to Florida. It adapts well to the laboratory environment, is quite hardy, and can be found in polluted as well as pristine areas. Initial work on populations from unpolluted areas of eastern Long Island (LI) (New York, USA) found that meHg at a concentration of 50 $\mu$g/l caused a variety of embryological malformations. These included craniofacial defects, manifested as an incomplete induction of the brain that allows progressive fusion of the eyecups, ultimately resulting in cyclopia or lack of brain and eye formation altogether (Fig. 1); cardiovascular defects, seen as progressive defects in heart development, including "tube hearts"; and various skeletal malformations, including vertebral bends and stunting (Weis and Weis 1977). Using indices of severity, which allow one to rank the craniofacial, cardiovascular,

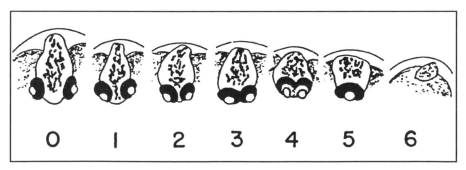

**FIG. 1.** Heads of 7-day-old mummichog embryos, demonstrating the craniofacial index for ranking anomalies: 0, normal; 3 synophthalmic; 5 cyclopic; 6 anophthalmic. From Weis et al. (1982a); Copyright American Society for Testing and Materials, reprinted with permission.

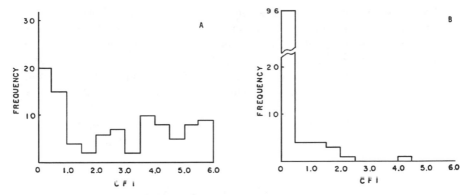

**FIG. 2.** Distribution of craniofacial index among treated batches of embryos from (a) Long Island and (b) Piles Creek. From Weis et al. (1981a); reprinted with permission from Springer–Verlag, New York Inc.

and skeletal defects in each embryo, we found considerable variation in responses from eggs of different females. Some females produced embryos that were highly resistant to this exposure, some females produced embryos that were very susceptible, and other females produced embryos of intermediate susceptibility (Fig. 2a). When we examined relationships of the susceptibility of a batch of eggs in order to see which traits of the male and female (length, weight, fin-ray count) or of the eggs themselves (number in the batch, stickiness, number of non-cleaving eggs) were more closely associated with the meHg tolerance, the trait with the strongest association was the number of dorsal fin rays of the female, a meristic character which has a high heritability (Tavc 1984; Leary et al. 1985). Fish with 10 fin rays produced no eggs that were unaffected, whereas fish with 11 or 12 fin rays produced eggs ranging from very tolerant (unaffected) to very susceptible. There was a weak correlation with female size in that larger fish produced more resistant eggs than smaller fish. Females tended to produce embryos with similar tolerance regardless of the male that fertilized the eggs. No correlations were seen with any of the traits of the male (Weis et al. 1982a). This finding should not be surprising because the critical period for the genesis of the craniofacial defects is during the second day of development (gastrulation) when paternal gene products are just beginning to be synthesized. Susceptible clutches of eggs accumulated higher amounts of meHg than tolerant ones, suggesting a role of the chorion in the accumulation of the toxicant that could result in the differential responses of the different batches of eggs. Other organisms have also been reported to use decreased uptake as a mechanism of metal tolerance (Foster 1977; Newman et al. 1994). Batches of eggs that were tolerant to meHg tended to be also tolerant to inorganic Hg (Weis et al. 1982a), but there was an inverse correlation with tolerance to lead. There was no cross-tolerance between meHg and tolerance to PCBs (Weis and Weis 1982b). Because eggs are not similarly tolerant to all xenobiotics or all metals, general chorionic permeability cannot be the mechanism, suggesting that the chorion of tolerant eggs may be selectively more impermeable to meHg.

The high variability in tolerance within the clean population allows part of it to be better able to withstand an influx of mercury contamination. Mitton and Koehn (1975) suggested that the polygynous mating system of this species allows for rapid gene frequency changes and, therefore, rapid evolutionary responses to a variable environment. Whether a population will be eliminated or will adapt to an introduced stress depends on the rapidity of onset and severity of the stress and the capacity of the population to adapt to it. The genetic diversity of *Fundulus* may allow this species to survive in polluted environments from which other species have been eliminated. Smith and Fujio (1982) noted that this species is one of the most genetically variable of all teleosts examined.

## Tolerance in a Contaminated Environment

We subsequently studied a population living in Piles Creek (PC), a contaminated estuary of the Newark Bay system (New Jersey, USA). This estuary has elevated levels of various heavy metals and organic contaminants in the sediments. The marsh around the creek is surrounded by industrial sites, a generating station, an oil refinery, a sewage treatment plant, and a major highway. Mercury levels in the sediments are 10–20 ppm, and lead levels are up to 3000 ppm. Mummichog livers contained about $0.5 \mu g\ g^{-1}$ whereas livers of the reference LI fish contained about one-tenth of this amount (Khan and Weis 1993a). Very few females in the PC population produced meHg-susceptible eggs, and most embryos were tolerant with respect to the production of embryological malformations, as seen in Fig. 2b (Weis et al. 1981a). Thus, the phenotypic variability seen in the LI population was not seen here, and there was much greater uniformity in the responses of different batches to meHg. This may reflect reduced population genetic diversity. As seen in LI, tolerant batches of eggs had less meHg uptake than susceptible ones. Magnusson and Ramel (1986) likewise observed that the level of methylmercury tolerance in *Drosophila* was correlated with the uptake of mercury, and Stuhlbacher et al. (1992) showed that cadmium tolerance in *Daphnia* was associated with differences in Cd uptake. The PC embryos developed more rapidly than in the LI ones, and more rapid development can also be a mechanism of resistance to teratogens. By passing through the sensitive critical stages more rapidly, the embryos can be protected from teratogenic effects of chemical exposure (Weis et al. 1981c). This is supported by data indicating that lower temperature, which slows down development, results in more severe defects from teratogens (Weis et al. 1981b; Munkittrick and Dixon 1988). PC females tended to have higher dorsal fin ray counts than in the LI populations, a trait that was associated in LI with greater tolerance of eggs. This suggests that selection for certain genotypes had taken place in PC.

## Rapid Change in Embryo Tolerance

Events in 1982 in Southampton LI demonstrated the rapidity with which a shift of embryonic tolerance could occur. That year there was an unusually large amount of

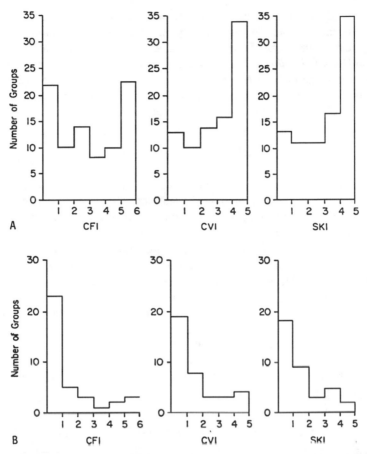

**FIG. 3.** (*a*) Distribution of number of clutches of meHg-treated embryos with each mean value of craniofacial index (CFI), cardiovascular index (CVI), and skeletal index (SKI) in Long Island in 1980. In all indices higher numbers represent more severe defects. (*b*) Distribution in 1982 after runoff from golf course. Reprinted from Weis and Weis (1984) with permission from Elsevier.

rainfall and associated runoff of pesticides in June from a golf course near one of our study sites. Subsequently, over 40% of the females produced nonviable eggs (which could not be tested for mercury tolerance), and the fish that produced viable eggs generally produced tolerant ones with respect to the craniofacial, cardiac, and skeletal defects (Fig. 3a and b; Weis and Weis 1984). Accompanying the shift in meHg tolerance was a trend toward increased dorsal fin rays of the females that produced viable eggs. We hypothesize that the influx of contaminants, including chlorinated hydrocarbon pesticides, caused the striking changes in reproductive success. The normal variability in the population allowed some fish to produce viable eggs, and these were the ones whose eggs were more resistant to meHg. These observations indicate that when a population is variable to begin with, a change in overall tolerance can happen very quickly. During the following summer, the tolerance to meHg returned to its usual heterogeneous state. The rapid change in tolerance may be similar to that observed by Rahel (1981) when he selected for zinc tolerance in a laboratory population of flagfish (*Jordanella floridae*). Increased resistance was seen after one generation, but continued selection for three more generations did not cause additional tolerance. He concluded that the exposure had culled out the susceptible individuals but did not cause genetic changes specifically related to zinc tolerance.

### Role of Female and Male

Reciprocal crosses were done between PC and LI fish to investigate tolerance of embryos having parents from different populations. These results supported earlier data showing that the tolerance of the embryos is determined solely by the female (Toppin et al. 1987). Munkittrick and Dixon (1988), studying a population of white suckers (*Catostomus commersoni*) from Cu-contaminated lakes, found enhanced Cu tolerance of early life history stages; these developed at a slightly faster rate than control eggs. The tolerance was not seen in larvae after the first feeding or in control eggs fertilized by sperm from the contaminated site. These authors concluded that this was evidence for a maternal factor contributing to the resistance. Hoare et al. (1995), studying copper resistance in embryos from a population of mussels, *Mytilus edulis*, from a polluted site, also found that the tolerance appeared to be maternally determined. Pierce and Wooten (1992), studying pH tolerance in wood frogs, *Rana sylvatica*, indicated that maternal factors (genetic or environmental) determined the variation in hatching success at low pH.

### Cytogenetic Effects of Methylmercury in Embryos

Methylmercury is known to cause cytogenetic effects in a variety of organisms. It was found to cause reduced mitotic count (as per method of Longwell and Hughes 1980) and an increased frequency of chromosomal aberrations in mummichog embryos (Perry et al. 1988). This study found that cytogenetic effects in *Fundulus* embryos correlated with teratogenic effects. Clutches that were susceptible to the teratogenic effects had a five-fold decrease in mitotic counts and a 2.5-fold increase in chromosomal aberrations when exposed to 50 $\mu$g/l meHg, whereas embryos from groups resistant to teratogenic effects had only a 1.5-fold decrease in mitotic count and two-fold increase in chromosomal aberrations (Table 1). Untreated embryos from the meHg susceptible groups, all morphologically normal, had more chromosomal anomalies than those of the resistant groups. A greater inherited tendency toward chromosomal aberrations may partly explain the greater toxicant effects in these groups. Furthermore, because a mechanism of embryonic tolerance is decreased uptake of the meHg, this reduced uptake should protect those embryos against the cytogenetic effects as well as teratogenic ones.

**TABLE 1.** *Effects of methylmercury on mean number of mitoses and chromosomal abnormalities in susceptible and resistant F. heteroclitus embryos*

| Condition | Treatment | n | Mean Number Mitoses | Total Percent Abnormalities[a] |
|-----------|-----------|---|---------------------|-------------------------------|
| Susceptible | Control | 10 | 212.0 ± 29.7 | 5.1 |
| | Experimental | 9 | 41.7 ± 12.8 | 13.1[b] |
| Resistant | Control | 35 | 174.8 ± 16.4 | 4.0 |
| | Experimental | 39 | 122.4 ± 11.5 | 7.4 |

$$^a\text{Total percent abnormal} = \frac{\text{Total number of abnormal divisions}}{\text{Total mitoses}}.$$

$^b\chi^2 = 96.6$, P. 0.05

(Reprinted from Perry et al, 1988, with the permission of Springer–Verlag, New York.)

## Search for a Physiological Component of Tolerance

An investigation into the possible role of metallothionein in conveying enhanced embryonic meHg tolerance revealed that newly deposited eggs lacked detectable levels of this metal-binding protein. Embryos did not synthesize it until shortly before hatching, which is well after the time period in which the abnormalities are produced (Weis 1984). However, Hg tolerant late embryos from LI had more metallothioneins than susceptible ones.

In order to further investigate the possibility that short-term physiological responses to Hg in the PC population might contribute to the enhanced embryonic tolerance to meHg, a number of experiments were performed (Weis et al. 1985):

1. LI fish were maintained in water with low concentrations (5 $\mu$g/l) of meHg or in water from PC in order to investigate whether subsequent batches of eggs would be more resistant to meHg. However, no subsequent batches of eggs were produced by these females.
2. PC females were maintained in clean water to see if the meHg tolerance of subsequent batches of eggs would be decreased. Tolerance was generally increased in later clutches, the opposite of what would be expected if tolerance were a short-term response to the environment. In addition, there was no correlation between the Hg content of a female's liver and the tolerance of the eggs she produced.
3. Tolerance of larvae that were pre-exposed to 20 $\mu$g/l meHg as embryos (a concentration that did not produce embryological malformations) was compared with that of larvae which had not been pre-exposed, using mortality ($LT_{50}$) as the measure of tolerance. The hypothesis was that the pre-exposed groups would be more tolerant if a short-term physiological mechanism were involved. The pre-exposed groups, however, were less resistant, rather than more (Weis and Weis 1983).
4. Adults that were pre-exposed to 5 and 10 $\mu$g/l meHg were tested in a fin regeneration assay and compared with adults that had not been pre-exposed (controls).

Although those pre-exposed to 5 $\mu$g/l meHg did not differ in fin regeneration rate from controls in their response to 50 $\mu$g/l during regeneration, those that had been pre-exposed to 10 $\mu$g/l, had a reduced regeneration rate compared to controls, indicating that they were less tolerant to meHg as a result of their pre-exposure. In summary, there is no evidence for a short-term component of meHg tolerance, and it is likely that the tolerance is genetic.

### Possible "Costs" of Tolerance

We found that eggs of fish from PC or from Newark Bay would not fertilize success-fully if stripped into full strength seawater (Bush and Weis 1983), but would fertilize only if the salinity was reduced. In contrast, eggs from the LI populations would fertilize successfully over a larger range of salinities from 10–30‰. If the PC eggs, stripped into 30‰ seawater, were transferred to 15‰ within one minute, successful fertilization could occur. Electron microscopy revealed that PC eggs stripped into 30‰ seawater underwent artificial activation; the micropyles became blocked with material extruded from the micropylar canal. This blockage closely resembled that seen in eggs after successful fertilization in 15‰ salinity water. This suggests that PC eggs are rapidly activated by contact with 30‰ seawater. Perhaps this population has adapted so narrowly to the specific conditions of its habitat (salinity ~15–20‰) that it has lost some of the euryplasticity present in the LI population. The development of tolerance to meHg, at the cost of reduced genetic variability in the population, may re-duce their ability to deal with natural stresses or other types of pollution. This may be a "cost" of the enhanced tolerance. Similarly, increased sensitivity to zinc was observed in a chromium-tolerant strain of *Daphnia magna* by Munzinger and Monicelli (1992).

### RESPONSES OF SPERM AND EGGS

The enhanced tolerance of PC embryos was also seen when eggs were exposed to meHg before fertilization, using fertilization success as the measure of response (Khan and Weis 1987b). After 20 minutes of exposure to various concentrations of meHg before fertilization with untreated sperm, PC eggs showed a higher $EC_{50}$ value than LI eggs (1.7 mg/l compared with 0.7 mg/l). In addition to the immediate effects, there were delayed effects on development. PC eggs that were exposed to 1.0 or 2.5 mg/l before fertilization and were successfully fertilized and placed in clean seawater for one week exhibited 5% and 7% malformations, respectively. In comparison, LI eggs that were successfully fertilized after exposure to 1.0 mg/l, exhibited 32% malforma-tions, and most of those that had been exposed to 2.5 mg/l before fertilization died during the first week of development. This indicates that the tolerance to meHg in the PC embryos is also present in the eggs of that population before fertilization. Scan-ning electron microscopy of eggs after exposure to meHg revealed that the decreased fertilization success was due to artificial activation caused by the rupture of cortical

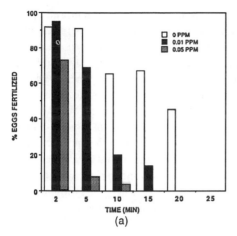
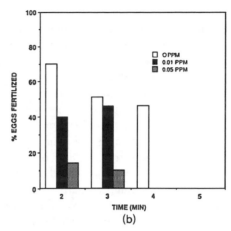

**FIG. 4.** (*a*) Effects of methylmercury on fertilization success of Piles Creek sperm. (*b*) Effects of methylmercury on fertilization success of Long Island sperm. Reprinted from Khan and Weis (1987a) with permission from Springer–Verlag.

vesicles and blockage of the micropyle, a reaction that occurred more frequently in the LI eggs than in the PC eggs (Khan and Weis 1993b).

Despite the finding that the male does not contribute to the tolerance of the embryos, sperm from PC males were considerably more tolerant to meHg exposure than sperm from LI males (Khan and Weis 1987a). LI sperm, after exposure to $10 \mu g/l$ meHg for 10 minutes, showed significantly reduced fertilization success due to reduced motility, whereas PC sperm were unaffected by this treatment (Fig. 4a and b). Untreated PC sperm also remained viable in clean seawater (15‰ S) for 20 minutes, whereas LI sperm remained viable for only 10 minutes. This might reflect lower permeability of the PC gametes, or may imply that the PC population devotes more energy to reproduction than the LI population. Similar results were seen by Munkittrick and Dixon (1988), who found that sperm from male white suckers from a metal-contaminated area performed better than sperm from control males in fertilization trials with control eggs.

## RESPONSES OF LARVAE AND JUVENILES

The studies previously discussed focused on tolerance to meHg by embryos and gametes. We also examined whether other life history stages also had elevated tolerance to meHg. Within the LI population, meHg-tolerant embryos tended to hatch into meHg-tolerant larvae (Weis and Weis 1982a), as measured by $LT_{50}$ (time to 50% mortality in 50 $\mu g/l$). However, in comparing tolerance of larvae from LI versus PC, no difference was found in acute response (Weis et al. 1987). Also, no difference was seen in the $LC_{50}$ for juveniles of the two populations (Khan and Weis 1987b). Thus, tolerance in the PC population appeared to be present only in the gametes and embryos. Because mechanisms contributing to embryonic tolerance are chorionic permeability and more rapid development, these are characteristics that do not apply to larvae. It is interesting to note that although larvae seemed to be the most sensitive stage in both populations (50 $\mu$g/l was generally lethal to larvae within one week, whereas many embryos were unaffected by this concentration), no tolerance was manifested at this stage when acute exposures were used. More recently, data have been obtained on behavioral effects (Zhou 1997) confirming that the PC larvae were not more resistant than the reference population. This agrees with the observations of Munkittrick and Dixon (1988) on early life stages of white suckers in a contaminated lake: tolerance to copper, which was manifested in earlier stages, was no longer present in larvae after the first feeding. Similarly, Moraitou-Apostolopoulou et al. (1982) found that enhanced cadmium tolerance in a contaminated population of *Palaemon elegans* was seen in younger but not older individuals.

## RESPONSES OF ADULTS

To compare adult tolerances of PC and LI populations, a fin regeneration assay was used. Regeneration requires an adequate nerve supply and involves wound healing, dedifferentiation and blastema formation, and growth. It can, therefore, be affected by toxicants that have different modes of action, and it encompasses developmental processes. Earlier work (Weis and Weis 1978) showed that exposure to 50 $\mu$g/l meHg retarded regeneration of partially amputated caudal fins. When the two populations were compared, it was found that the baseline (control) values were very different. PC fish regenerated very slowly compared to LI fish. The control values for PC were even lower than meHg-treated LI fish. Although meHg exposure retarded both groups to about the same degree, there was significant mortality in the PC but not the LI population, indicating decreased tolerance to the meHg exposure (Fig. 5; Weis et al. 1987). Thus, PC adults showed signs of stress, slower growth, and decreased tolerance, rather than increased tolerance. Stress caused by pollutants in PC may have reduced the energy available for regenerative growth. Exposure of regenerating adults to 30 $\mu$g/l produced enhanced growth (hormesis) in the LI but not the PC population. Hormesis, or "sufficient challenge," is believed to be an overcompensation by homeostatic regulatory control mechanisms to a low concentration of a toxicant (Stebbing 1981). PC fish, already stressed in their natural environment, appear to be unable to muster this response. Because the ability to counteract toxicants can be considered a measure of health, these data further support the idea that the PC adults are in poor health. It is likely that the poor health of the adults is a result of living in a polluted environment rather than a lowered genetic fitness, although the latter is possible. In contrast, adult mummichogs living near a site contaminated with high levels of PAHs did demonstrate increased resistance to PAHs (van Veld and Westbrook 1995). Similarly, Prince and Cooper (1995) have also found that *Fundulus* from Newark Bay are more tolerant to dioxin, both as embryos and as adults.

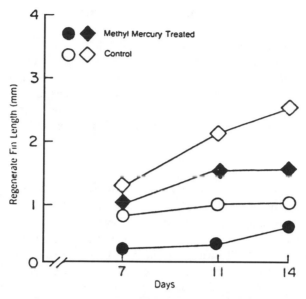

**FIG. 5.** Caudal fin regeneration in mummichogs from Piles Creek (circles) and Long Island (diamonds) under control conditions (open) or treated with 50 μg/l methylmercury (filled). From Weis et al. (1987); reprinted with permission from Krieger Publishing Company.

## LIFE HISTORY ALTERATIONS
## IN A CONTAMINATED ENVIRONMENT

In examining the adults, it was noted that they were generally smaller than those from reference populations. When we investigated size/age relationships, it became clear that the females from PC are comparable to reference fish up to one year of age, but after one year do not grow as well or live as long as reference fish (Toppin et al. 1987). We also observed that the condition index of the PC fish was significantly lower than that of reference LI fish (Weis and Khan 1991). At the reference sites, occasional four-year-old fish were found, but no PC fish beyond three years of age were found. However, females in the field became reproductive at a smaller size and younger age (Toppin et al. 1987) and fertility was high. They can thus produce the next generation of tolerant embryos and perpetuate the population. The earlier maturation and decreased longevity is similar to the findings of MacFarlane and Franzin (1978) on white suckers from a metal-contaminated lake. Tranvik et al. (1993) likewise found that *Collembola* species from polluted sites had higher overall reproduction rates than those from reference sites. Similarly, Donker et al. (1993), studying metal-adapted populations of the terrestrial isopod *Porcellio scaber*, found earlier reproduction and increased reproductive allocation, compared with reference populations. They concluded that selection for early reproduction and increased reproductive allocation had taken place. The longer viability we noted of the PC gametes in seawater (Khan and Weis 1987a) may further indicate that PC fish devote more energy to reproduction.

## BEHAVIORAL ALTERATIONS

More recently, we have acquired insight into possible reasons for decreased growth of the PC fish. Using fish from either Tuckerton, New Jersey (TT) or from LI as the reference population, we have found that PC fish are slower to capture prey (Weis and Khan 1991; Smith et al. 1995). We have also found that exposure to meHg for as little as one week can also reduce the prey capture rate of the PC population (Weis and Khan 1990). Gut content analysis has shown that in the field, PC fish eat more detritus and less live food than TT fish, despite the healthy population of grass shrimp, a favored prey item, at PC (Smith and Weis 1997). Other workers have noted that although mummichogs eat detritus, it does not provide them with nutrition (Prinslow et al. 1974). Thus, PC fish, by eating more detritus and less live food, have a poor diet, which can contribute to their reduced growth and condition.

The polluted environment in PC seems to be largely responsible for the impaired prey capture ability of the PC fish. When TT fish were kept for two months in PC conditions (aquarium with PC water, sediment, and shrimp to eat), their prey capture ability declined to that of PC fish, and their brain mercury level increased to that of PC fish. When PC fish were maintained in clean water to see if their predatory behavior would improve, the improvement was very slight, and their brain mercury level did not decline (Smith and Weis 1997). Analysis of videotapes revealed that the poor prey capture ability of the PC fish was due primarily to reduced attempts to capture the prey (grass shrimp, *Palaemonetes pugio*), reflecting their generally slow, sluggish activity. This sluggishness also resulted in increased vulnerability to predation by blue crabs in laboratory aquaria. If fish with greater levels of contaminants in their systems are easier for predators to capture, this is a way for greater amounts of toxicants to move through the food web. The altered behavior may, in turn, be caused by reduced levels of the neurotransmitter serotonin in their brains (Smith et al. 1995). Thus, a feasible scenario is that continued exposure to neurotoxicants in the environment may alter neurotransmitters, which impairs prey capture behavior and results in decreased diet quality, altered growth, size-structure, and life history. Thus, connections may be made from the cell/biochemical level (neurotransmitters) to the individual (behavior) to the population (reduced growth and longevity).

However, the behavioral "defects" in the PC population may not be completely due to neurotoxic chemicals. There seem to be some inherent (genetic?) factors that contribute to activity level and prey capture ability. When PC larvae are a few days old, they exhibit greater activity and prey capture rates than TT larvae. However, by the time they are two weeks old, their activity and prey capture rates have dropped and are lower than TT (Zhou 1997). These larvae were kept in the laboratory in clean water, and thus inherent factors appear to be responsible for the behavioral differences. It is also possible that residual chemicals remaining in the larvae from oogenesis in a contaminated environment could be exhibiting delayed effects and altering the larval behavior. However, one would expect the effects to be manifested in the early larval stages rather than later. We do not as yet have an explanation for the changes in larval behavior in fish raised in clean water.

## INVESTIGATIONS INTO POPULATION GENETICS

### Morphology

It has been known for some time that there are separate northern and southern "races" or subspecies of *F. heteroclitus* and that the intergradation zone is in New Jersey. This can confound efforts to relate resistance to genetics. Morphological studies (Able and Felley 1986) indicated that fish from northern localities have more predorsal scales, lateral scales, and gill rakers and that females from northern populations tend to have shorter anal sheaths than females from southern locations. The variation in adult morphology is concordant with egg morphology. A striking difference in shape and number of chorionic fibrils was reported by Brummett (1966), when embryos from Woods Hole, Massachusetts, and Beaufort, North Carolina, were compared. Woods Hole females produced eggs with long sticky fibrils that the southern eggs lacked. Electron microscopy revealed that chorionic fibrils of the northern population are very long, thick, and sparsely distributed, whereas most of the fibrils of the southern population are shorter, thinner, and very densely distributed. The number of oil droplets per egg increased clinally from Maine to Florida. The pattern of variation in the external morphology of the chorion, described by Brummett and Dumont (1981), had a very steep cline with the northern pattern observed to the north, and the southern pattern to the south of Raritan Bay, New Jersey (Morin and Able 1983).

Based on this large set of meristic, morphological, and genetic data, Able and Felley (1986) suggested that, "These morphs should be treated as separate taxa, with *F.h. heteroclitus* occurring along the east coast from New Jersey south to Florida, including lower Chesapeake and Delaware bays and *F.h. macrolepidotus* distributed from Connecticut north to Newfoundland, but the lack of morphological diagnostic characters makes field recognition of the subspecies very difficult."

### Allozymes

Studying a Long Island Sound population, Mitton and Koehn (1975) found that over 50% of 25 loci scored were polymorphic with an average heterozygosity of 18.5% per locus per individual. Powers et al. (1986) extended their study and found 16 of 45 loci to be polymorphic. Examination of the geographical distribution uncovered significant directional changes in gene frequency (i.e., clines) and degree of heterozygosity with latitude at some loci.

Lactate dehydrogenase-B (*Ldh*-B) has two predominant alleles: a fast allele (a) fixed at 1.00 at Savannah, GA (latitude 32.08N), whose frequency slowly decreases northward to 0.627 at Point Pleasant, New Jersey (40.09N). Then, within less then a degree of latitude, it drops down to 0.25 at Stony Brook, New York (40.93N), being replaced by the (b) allele, and keeps decreasing slowly to 0.00 at Shediac, New Brunswick, Canada (46.25N) (Powers and Place 1978; Ropson et al. 1990).

Malate dehydrogenase-A (*Mdh*-A) has a significantly sharper cline in the same geographical area, being almost fixed to the north and south from the New York/New Jersey Harbor for (a) and (b) alleles, respectively. Between Point Pleasant and Stony Brook there is a shift from 0.020 to 0.952 for the northern allele (Powers and Place 1978).

Four alleles are known for glucosephosphate isomerase-B (*Gpi*-B); two (b and c) being common and two others (a and d) being present in fewer than 5% of the individuals scored (Place and Powers 1978). The (*Gpi*-B$^c$) allele frequency reaches 1.00 at Bar Harbor, Maine (44.43N), but few individuals with (b) occur further north (Ropson et al 1990); it decreases to the south to 0.895 at Stony Brook. From Point Pleasant to the southern edge of the species distribution it fluctuates between 0.654 and 0.234 without a clear clinal pattern (Powers and Place 1978).

Mannosephosphate isomerase-A (*Mpi*-A) is variable only within the northern subspecies populations. It has four alleles (a,b,c,d), two of which are rare (less than 5 (a) and 8(c)%). Being about 10% within all *F.h. heteroclitus* populations, allele (d) shows significant variation (0.172 to 1.00) in *F.h. macrolepidotus* north from Long Island. The sharpest shift in allele frequency (0.286–0.890) occurs between populations on different sides of Cape Cod (Ropson et al. 1990).

Hexose-6-phosphate dehydrogenase-A (H6pdh-A), like the previous enzyme, has a major shift in allele (c) gene frequency from the south to north side of Cape Cod (from 0.415 to 0.857) and further north it becomes fixed at 1.00. In *F.h. heteroclitus*, it fluctuates smoothly between 0.06 and 0.294 without any clinal patterns.

Aspartate aminotransferase-A (*Aat*-A) has substantial variation only in the middle of the range of *F.h. macrolepidotus* at Wiskasset, Maine (frequency of rare allele (b) = 0.358) and Salisbury, Massachusetts (0.185). In other populations this allele frequency is never more than 2%. The second locus, *Aat*-B, has some degree of polymorphism in all populations studied (Ropson et al. 1990).

Esterase-B (*Est*-B), Isocitrate dehydrogenase -A (*Idh*-A), *Idh*-B, Fumarase-A (*Fum*-A), and Phosphoglucomutase-B (*Pgm*-B), and *Pgm*-A have some degree of polymorphism in all populations studied except the most northern, where the most common allele is fixed at 1.00. Only Acid Phosphatase-A (*Ap*-A) and Esterase-C (*Est*-C) remain at the same level of polymorphism through all the populations studied (Powers and Place 1978; Powers et al. 1986; Brown et al. 1988; Ropson et al. 1990).

## Mitochondrial DNA

A large phylogenetic distance has been revealed between the two subspecies when the mitochondrial DNA restriction fragment length polymorphism (RFLP) technique has been applied. Specimens from Maine to northern New Jersey were found to possess a very homogeneous "north" haplotype, whereas all specimens from Sapelo Island, Georgia, and southern New Jersey were found to have the "south" one (Gonzales-Villasenor and Powers 1990).

Molecular phylogenies using mitochondrial DNA (cytochrome b gene) and nuclear alleles of the lactate dehydrogenase B locus have been studied (Bernardi et al 1993).

Both mitochondrial DNA and lactate dehydrogenase alleles show a clear separation between the northern individuals (from Nova Scotia and Maine) and the southern ones (from Georgia and Florida).

## Local Population Genetics

An intensive survey of the New York/New Jersey populations was done in order to ascertain the pattern of gene frequency distribution between sites previously reported as "southern" (Manasquan River, New Jersey, predominantly $Mdh$-$A^b$) and "northern" (Stony Brook, fixed at $Mdh$-$A^a$) and to see if the allozyme clines were concordant with the mtDNA (Mugue 1996). Forty marshes were sampled along the Atlantic Coast from southern New Jersey through the New York/New Jersey Harbor estuaries, including the Hudson and Hackensack rivers and the highly polluted estuaries of the Newark Bay system (including PC) to Connecticut and LI (Fig. 6).

Four loci ($Mdh$-A, $Ldh$-B, $Gpm$-A, and $Gpi$-B) were chosen for the routine analysis in all populations. These four loci contribute much in mummichog polymorphism in this geographical area (Powers et al. 1986) and include the presumably diagnostic locus ($Mdh$-A). Protein electrophoresis was conducted using standard techniques. Genetic interpretation of electrophoretic patterns is based on previous genetic analyses (Powers et al. 1986). Population genetic analysis was performed on the BIOSYS_I program (Swofford and Selander 1981).

Frequencies of $Mdh$-A, $Ldh$-B, $Pgm$-A, and $Gpi$-B loci for all populations studied are presented in Table 2 and Figs. 6 and 7. The most steep and concordant clines occur between Wreck Pond (WP) and Shark River (Belmar, BM) estuaries, where both mtDNA and $Mdh$-A allelic frequencies shift significantly within 5 km and clearly indicate the current subspecies' boundary. All Atlantic shore populations from Tuckerton (TT) to the Manasquan River (Gull Island, GI) are fixed or nearly fixed in $Mdh$-$A^b$, making this allele a reliable marker for $F.h.$ $heteroclitus$. The frequency of another allele of this locus, $Mdh$-$A^a$, increases from 6% in the Manasquan River (GI) to 22% at Wreck Pond (WP). From BM (the Shark River at Belmar, 65.7%) to the North $Mdh$-$A^a$ became the major allele. Northward from Belmar to the lower parts of the Hudson and the Hackensack River estuaries and western LI Sound, $Mdh$-$A^a$ remains at average frequency of 60%, presented as a "plateau" on the plot frequency versus distance. After this plateau, the second abrupt shift in gene frequencies can be seen where $Mdh$-$A^a$ frequency of 60% quickly approaches 90–100% at more upstream or northern locations.

MtDNA haplotype frequencies from all sites studied are shown in Table 3 and Figs. 6 and 7. Populations at Tuckerton (TT) consist of fish with only mtDNA(S) haplotypes, and all populations of Raritan Bay (PI, PM, and CC), the Hackensack (NK, PR, BC, VL, HG, and CB) and Hudson (PT, SP, BR, CW, MB, and TB) rivers, and Long Island Sound (CS, CP, DI, NP, CM, SM, and WM) consist of fish that possess exclusively the northern haplotype (mtDNA(N)). In populations from Barnegat bay (GL) to the Shark River (BM), both haplotypes are represented in different proportions.

**TABLE 2.** *Mdh-A, Ldh-B, Pgm-A, and Pgl-B frequencies at the sites studied*

| Site | km | MDH-A | | | | | LDH-B | | | | | PGM-A | | | | | | PGI-A | | | | | | | |
|---|---|---|---|---|---|---|---|---|---|---|---|---|---|---|---|---|---|---|---|---|---|---|---|---|---|
| | | (N) | A | B | C | SE(A) | (N) | A | B | C | SE(A) | (N) | A | B | C | D | SE(B) | (N) | A | B | C | D | E | SE(B) |
| TT | 0 | 65 | 0.000 | 1.000 | 0.000 | 0.000 | 65 | 0.569 | 0.431 | 0.000 | 0.043 | 65 | 0.008 | 0.800 | 0.192 | 0.000 | 0.035 | 65 | 0.000 | 0.415 | 0.585 | 0.000 | 0.000 | 0.043 |
| OC3 | 42 | 20 | 0.050 | 0.950 | 0.000 | 0.034 | 20 | 0.650 | 0.350 | 0.000 | 0.075 | 20 | 0.000 | 0.950 | 0.050 | 0.000 | 0.034 | 20 | 0.000 | 0.400 | 0.600 | 0.000 | 0.000 | 0.077 |
| OC1 | 43 | 20 | 0.025 | 0.975 | 0.000 | 0.025 | 20 | 0.700 | 0.300 | 0.000 | 0.072 | 20 | 0.000 | 0.875 | 0.125 | 0.000 | 0.052 | 20 | 0.000 | 0.475 | 0.525 | 0.000 | 0.000 | 0.079 |
| FR | 45 | 29 | 0.000 | 1.000 | 0.000 | 0.000 | 29 | 0.655 | 0.328 | 0.017 | 0.062 | 29 | 0.000 | 0.793 | 0.207 | 0.000 | 0.053 | 29 | 0.000 | 0.328 | 0.672 | 0.000 | 0.000 | 0.062 |
| GL | 55 | 237 | 0.019 | 0.979 | 0.002 | 0.006 | 213 | 0.667 | 0.333 | 0.000 | 0.023 | 60 | 0.008 | 0.817 | 0.175 | 0.000 | 0.035 | 181 | 0.006 | 0.406 | 0.583 | 0.006 | 0.000 | 0.026 |
| GI | 78 | 42 | 0.060 | 0.917 | 0.024 | 0.026 | 42 | 0.452 | 0.548 | 0.000 | 0.054 | 10 | 0.000 | 0.700 | 0.300 | 0.000 | 0.102 | 42 | 0.000 | 0.262 | 0.738 | 0.000 | 0.000 | 0.048 |
| WP | 83 | 25 | 0.220 | 0.780 | 0.000 | 0.059 | 25 | 0.680 | 0.320 | 0.000 | 0.066 | 10 | 0.000 | 0.800 | 0.200 | 0.000 | 0.089 | 25 | 0.000 | 0.400 | 0.600 | 0.000 | 0.000 | 0.069 |
| BM | 88 | 89 | 0.567 | 0.433 | 0.000 | 0.037 | 87 | 0.454 | 0.546 | 0.000 | 0.038 | 74 | 0.007 | 0.872 | 0.122 | 0.000 | 0.027 | 89 | 0.000 | 0.292 | 0.708 | 0.000 | 0.000 | 0.034 |
| PI | 135 | 25 | 0.540 | 0.460 | 0.000 | 0.070 | 25 | 0.540 | 0.460 | 0.000 | 0.070 | 25 | 0.060 | 0.860 | 0.080 | 0.000 | 0.049 | — | — | — | — | — | — | — |
| PM | 145 | 20 | 0.600 | 0.400 | 0.000 | 0.077 | 20 | 0.350 | 0.650 | 0.000 | 0.075 | 20 | 0.050 | 0.850 | 0.100 | 0.000 | 0.056 | 20 | 0.000 | 0.200 | 0.800 | 0.000 | 0.000 | 0.063 |
| CC | 164 | 15 | 0.500 | 0.500 | 0.000 | 0.091 | 15 | 0.333 | 0.667 | 0.000 | 0.086 | 15 | 0.000 | 0.900 | 0.100 | 0.000 | 0.055 | 15 | 0.000 | 0.133 | 0.867 | 0.000 | 0.000 | 0.062 |
| PC | 182 | 32 | 0.672 | 0.328 | 0.000 | 0.059 | 32 | 0.516 | 0.484 | 0.000 | 0.062 | 32 | 0.000 | 0.922 | 0.078 | 0.000 | 0.034 | 32 | 0.000 | 0.250 | 0.750 | 0.000 | 0.000 | 0.054 |
| NW | 269 | 29 | 0.897 | 0.103 | 0.000 | 0.040 | 29 | 0.310 | 0.690 | 0.000 | 0.061 | 19 | 0.000 | 1.000 | 0.000 | 0.000 | 0.000 | 29 | 0.000 | 0.155 | 0.845 | 0.000 | 0.000 | 0.048 |
| NH | 318 | 45 | 0.967 | 0.033 | 0.000 | 0.019 | 45 | 0.156 | 0.844 | 0.000 | 0.038 | 44 | 0.000 | 0.989 | 0.011 | 0.000 | 0.011 | 44 | 0.000 | 0.102 | 0.886 | 0.011 | 0.000 | 0.032 |
| SD | 75 | 35 | 0.128 | 0.872 | 0.000 | 0.040 | 35 | 0.543 | 0.457 | 0.000 | 0.060 | 34 | 0.000 | 0.735 | 0.265 | 0.000 | 0.054 | 35 | 0.000 | 0.429 | 0.571 | 0.000 | 0.000 | 0.059 |
| SC | 85 | 10 | 0.100 | 0.900 | 0.000 | 0.067 | 10 | 0.500 | 0.500 | 0.000 | 0.112 | 8 | 0.000 | 0.875 | 0.125 | 0.000 | 0.083 | 10 | 0.000 | 0.450 | 0.450 | 0.100 | 0.000 | 0.111 |
| HV | 88 | 10 | 0.150 | 0.850 | 0.000 | 0.080 | 10 | 0.550 | 0.450 | 0.000 | 0.111 | 10 | 0.000 | 0.900 | 0.100 | 0.000 | 0.067 | 10 | 0.000 | 0.400 | 0.600 | 0.000 | 0.000 | 0.110 |
| PL | 236 | 20 | 0.650 | 0.350 | 0.000 | 0.075 | 20 | 0.450 | 0.550 | 0.000 | 0.079 | 20 | 0.000 | 0.975 | 0.025 | 0.000 | 0.025 | 20 | 0.000 | 0.075 | 0.925 | 0.000 | 0.000 | 0.042 |

| | N | | | | | | N | | | | | | N | | | | | | | N | | | | | |
|---|---|---|---|---|---|---|---|---|---|---|---|---|---|---|---|---|---|---|---|---|---|---|---|---|---|---|
| SP | 257 | 26 | 0.673 | 0.327 | 0.000 | 0.065 | 26 | 0.269 | 0.731 | 0.000 | 0.061 | 16 | 0.000 | 0.969 | 0.031 | 0.000 | 0.031 | 16 | 0.000 | 0.063 | 0.938 | 0.000 | 0.043 |
| BR | 269 | 28 | 0.696 | 0.304 | 0.000 | 0.061 | 28 | 0.321 | 0.679 | 0.000 | 0.062 | 10 | 0.000 | 1.000 | 0.000 | 0.000 | 0.000 | 14 | 0.000 | 0.071 | 0.929 | 0.000 | 0.049 |
| CW | 287 | 21 | 0.857 | 0.143 | 0.000 | 0.054 | 21 | 0.357 | 0.643 | 0.000 | 0.074 | 21 | 0.000 | 1.000 | 0.000 | 0.000 | 0.000 | 21 | 0.000 | 0.143 | 0.857 | 0.000 | 0.054 |
| MB1 | 302 | 32 | 0.969 | 0.031 | 0.000 | 0.022 | 32 | 0.203 | 0.797 | 0.000 | 0.050 | 32 | 0.000 | 1.000 | 0.000 | 0.000 | 0.000 | 32 | 0.000 | 0.031 | 0.969 | 0.000 | 0.022 |
| TBN | 343 | 22 | 0.977 | 0.023 | 0.000 | 0.023 | 22 | 0.114 | 0.886 | 0.000 | 0.048 | 22 | 0.000 | 1.000 | 0.000 | 0.000 | 0.000 | 22 | 0.000 | 0.000 | 1.000 | 0.000 | 0.000 |
| NK | 192 | 16 | 0.531 | 0.469 | 0.000 | 0.088 | 16 | 0.531 | 0.469 | 0.000 | 0.088 | 16 | 0.031 | 0.938 | 0.031 | 0.000 | 0.043 | 16 | 0.000 | 0.031 | 0.969 | 0.000 | 0.031 |
| PR | 202 | 19 | 0.526 | 0.474 | 0.000 | 0.081 | 20 | 0.350 | 0.650 | 0.000 | 0.075 | 20 | 0.000 | 0.925 | 0.075 | 0.000 | 0.042 | 16 | 0.000 | 0.219 | 0.781 | 0.000 | 0.073 |
| BC | 208 | 24 | 0.646 | 0.354 | 0.000 | 0.069 | 24 | 0.500 | 0.500 | 0.000 | 0.072 | 24 | 0.042 | 0.938 | 0.021 | 0.000 | 0.035 | 24 | 0.000 | 0.167 | 0.833 | 0.000 | 0.054 |
| VL | 210 | 78 | 0.538 | 0.462 | 0.000 | 0.040 | 78 | 0.353 | 0.641 | 0.006 | 0.038 | 33 | 0.015 | 0.894 | 0.091 | 0.000 | 0.038 | 34 | 0.000 | 0.206 | 0.794 | 0.000 | 0.049 |
| HG | 219 | 30 | 0.883 | 0.117 | 0.000 | 0.041 | 30 | 0.317 | 0.683 | 0.000 | 0.060 | 25 | 0.020 | 0.940 | 0.020 | 0.020 | 0.034 | 30 | 0.000 | 0.217 | 0.783 | 0.000 | 0.053 |
| CB | 223 | 34 | 0.882 | 0.118 | 0.000 | 0.039 | 34 | 0.309 | 0.691 | 0.000 | 0.056 | 9 | 0.000 | 0.944 | 0.056 | 0.000 | 0.054 | 20 | 0.000 | 0.250 | 0.700 | 0.025 | 0.068 |
| CS | 251 | 50 | 0.790 | 0.210 | 0.000 | 0.041 | 62 | 0.169 | 0.831 | 0.000 | 0.034 | 53 | 0.000 | 0.953 | 0.047 | 0.000 | 0.021 | 61 | 0.008 | 0.115 | 0.877 | 0.000 | 0.029 |
| CP | 275 | 30 | 0.817 | 0.183 | 0.000 | 0.050 | 30 | 0.150 | 0.850 | 0.000 | 0.046 | 30 | 0.000 | 0.983 | 0.017 | 0.000 | 0.017 | 30 | 0.033 | 0.150 | 0.817 | 0.000 | 0.046 |
| DI | 285 | 22 | 0.818 | 0.182 | 0.000 | 0.058 | 27 | 0.185 | 0.815 | 0.000 | 0.053 | 27 | 0.000 | 0.981 | 0.019 | 0.000 | 0.019 | 27 | 0.019 | 0.093 | 0.889 | 0.000 | 0.040 |
| NP | 295 | 36 | 0.722 | 0.278 | 0.000 | 0.053 | 36 | 0.292 | 0.708 | 0.000 | 0.054 | 36 | 0.000 | 0.972 | 0.028 | 0.000 | 0.015 | 36 | 0.000 | 0.153 | 0.847 | 0.000 | 0.042 |
| CM | 297 | 57 | 0.772 | 0.228 | 0.000 | 0.039 | 57 | 0.237 | 0.763 | 0.000 | 0.040 | 57 | 0.000 | 0.974 | 0.026 | 0.000 | 0.015 | 57 | 0.000 | 0.079 | 0.921 | 0.000 | 0.025 |
| SM | 305 | 28 | 0.893 | 0.107 | 0.000 | 0.041 | 28 | 0.304 | 0.696 | 0.000 | 0.061 | 28 | 0.000 | 0.929 | 0.071 | 0.000 | 0.034 | 28 | 0.000 | 0.089 | 0.911 | 0.000 | 0.038 |
| WM | 313 | 30 | 0.933 | 0.067 | 0.000 | 0.032 | 30 | 0.167 | 0.833 | 0.000 | 0.048 | 30 | 0.000 | 0.900 | 0.100 | 0.000 | 0.035 | 30 | 0.000 | 0.133 | 0.867 | 0.000 | 0.044 |

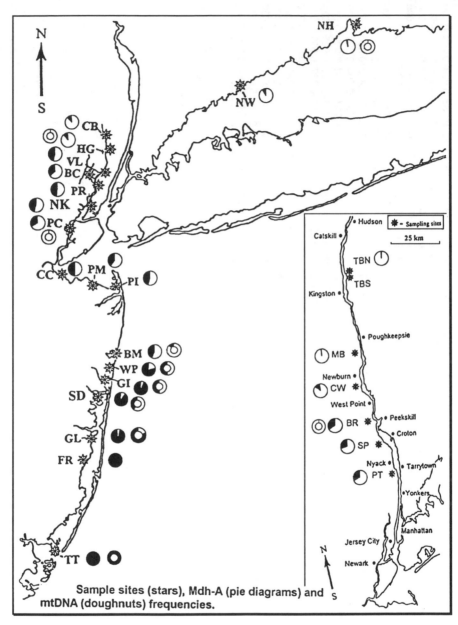

**FIG. 6.** Sample sites in NJ and NY (stars), *Mdh*-A frequencies (pie diagrams), and mtDNA frequencies (doughnuts).

**TABLE 3.** *MtDNA haplotype frequencies along the NJ shore*

| Site | km | (N) | N | S | SE |
|------|-----|-----|-------|-------|-------|
| | | | mtDNA | | |
| TT | 0 | 5 | 0.000 | 1.000 | 0.000 |
| GL | 55 | 5 | 0.200 | 0.400 | 0.179 |
| GI | 78 | 20 | 0.500 | 0.500 | 0.112 |
| WP | 83 | 19 | 0.473 | 0.527 | 0.115 |
| BM | 88 | 29 | 0.870 | 0.130 | 0.062 |
| PC | 182 | 5 | 1.000 | 0.000 | 0.000 |
| NH | 318 | 5 | 1.000 | 0.000 | 0.000 |
| SD | 73 | 10 | 0.600 | 0.400 | 0.155 |
| SC | 85 | 7 | 0.428 | 0.572 | 0.187 |
| HV | 88 | 8 | 0.625 | 0.375 | 0.171 |
| BR | 269 | 9 | 1.000 | 0.000 | 0.000 |
| HG | 219 | 5 | 1.000 | 0.000 | 0.000 |

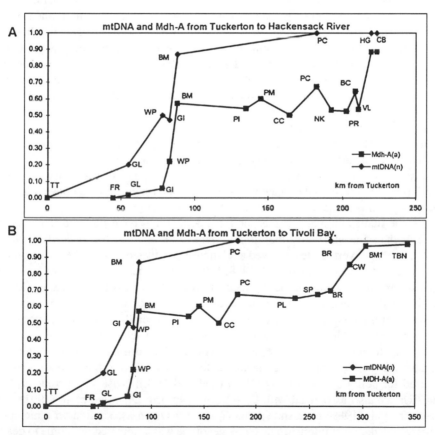

**FIG. 7.** (*a*) Mdh-A$^a$ and mtDNA$^n$ frequency from Tuckerton (TT, 0 km) to Coles Brook on the Hackensack River. (*b*) Mdh-A$^a$ and mtDNA$^n$ frequency from Tuckerton to Tivoli Bay (Hudson River).

Population genetic analysis thus reveals three distinctive groups of mummichog populations that inhabit the area covered in our study. Populations from Tuckerton to Wreck Pond have a southern mtDNA haplotype and $Mdh$-A$^b$ allele. According to Able and Felley (1986), these populations represent *F.h. heteroclitus*. In the upper river estuaries and in Connecticut sites mummichogs have a very high frequency of $Mdh$-A$^a$ and a northern mtDNA haplotype, and therefore represent the northern counterpart—*F.h. macrolepidotus*. In the area between these two genetically distinct subspecies, cluster analysis of allozyme loci frequencies reveals a third group of populations. With northern type of mtDNA (80% at BM and 100% in other sites), these populations have transitional frequencies of $Mdh$-A alleles, and, therefore, they are highly polymorphic at this locus (50–75% for $Mdh$-A$^a$).

Three highly polluted areas are among our sample sites: PC, Berry's Creek (BC), near a former mercury recycling facility where mercury content in the soil was recently as high as 1%, and Newark Bay (NK). All of these locations are occupied by the intermediate form, but because these sites lie within its current geographical range and fish have genetic structures in keeping with their geographical location, one cannot ascertain from the genes examined whether there have been any genetic changes associated with the tolerance to contaminants. However, some data may suggest that selection has occurred. DiMichele and Powers (1984) showed that developmental rate is faster in *Fundulus* embryos with the LDH-B$^a$B$^a$ genotype. This is the genotype that predominates in the southern populations. This trait would presumably convey increased embryonic tolerance. The developmental rate of PC embryos (intermediate form) is faster than that of both TT (southern) and LI (northern) embryos. It can be seen in Table 2 that in the three polluted sites, PC, NK, and BC, the proportion of LDH-B$^a$B$^a$ is about 0.5, which is somewhat higher than at nearby sites, where it tends to be about 0.3. This might indicate selection for embryonic resistance via faster development at these contaminated sites. On the other hand, TT fish have a higher frequency of this allele than PC fish and still develop more slowly.

In some studies on other species, polluted populations have shown differences in some of the same allozymes that do not seem to be selected for in our fish. Nevo et al., studying invertebrates, showed a relationship between $Pgm$ and $Pgi$ allozymes and mercury tolerance (Nevo et al. 1978, 1981, 1984). Mercury was found to select against rare homozygotes at the $Pgi$ locus in five different gastropod species (Lavie and Nevo 1986), and the gene frequencies in natural populations of these species showed that the mercury-resistant genotypes were especially abundant at sites with elevated levels of this contaminant (Nevo et al. 1984). Nevo et al. (1981) found differential tolerance of certain $Pgm$ genotypes of the shrimp *Palaemon elegans* to mercury, suggesting that they are adaptive; this was verified in the field so it could be used for pollution monitoring (Nevo et al. 1984). Likewise in gastropods, $Pgi$ and $Pgm$ are also associated with Cd tolerance. In fish, Diamond et al. (1991) and Newman et al. (1989) have shown that $Gpi$ may have an association with mercury resistance in the mosquitofish. The $Gpi$-2 genotype was associated with time to death in acute exposures of *Gambusia holbrooki* to inorganic mercury (Heagler et al. 1993). In particular, $Gpi$-2$^{38}$ homozygotes had a lower survivorship following Hg exposure than other genotypes at this locus. However, there are major differences between acute exposures in the lab and a field population that is confronted with a variety of pollutant and natural stresses. Nevertheless, they found that a field population in a mercury-contaminated stream did have altered frequency of a *Gpi-2* allele compared to fish from an adjacent uncontaminated site. On the other hand, analysis of Hg tolerance in fish from various natural populations showed no fitness differences among genotypes at the *Gpi-2* locus (Diamond et al. 1991), and the fish from the exposed population did not manifest increased tolerance over the control population, indicating that adaptation to Hg has not occurred. Mulvey et al. (1995) admitted that tolerance may not be directly associated with *Gpi* but that the genes for tolerance may be linked to the *Gpi-2* locus. Herbert and Luiker (1996) have stated that the usefulness of allozymes as indicators of pollutant exposure requires the identification of loci demonstrating consistent responses to the particular contaminant. There are a number of other cases in which genetic differentiation between contaminated and reference populations has not been seen, so that allozymes are not always useful as biomonitors of pollution (Diamond et al. 1991). The study of allozyme variability, after all, involves investigation of a very limited subset of proteins coded by a very limited subset of structural genes. A study of resistance to mercury in the mosquitofish *Gambusia holbrooki* did not find heritability for resistance (Lee et al. 1992).

Studies have generally not been able to separate direct effects on the locus being studied from indirect effects resulting from selection at linked loci. It is likely that the actual genes conveying resistance are not the genes for the allozymes studied, but other genes with which they are linked. As Forbes (1996) has stated, "Although changes in the genetic structure of populations have been correlated with pollution levels, the functional basis for such correlations is often obscure." It is clear that exposure to toxicants may exert a stress which can modify population genotype frequencies. All the evidence from our studies of meHg tolerance in the mummichogs points to a genetic rather than physiological basis of the tolerance. However, we have not yet found any particular genes associated with or responsible for this tolerance.

## REFERENCES

Able, K. W., and J. D. Felley. 1986. Geographical variation in *Fundulus heteroclitus*: Test for concordance between egg and adult morphologies. *American Zoologist* 26:145–157.

Benson, W. H., and W. J. Birge. 1985. Heavy metal tolerance and metallothionein induction in fathead minnows: Results from field and laboratory investigations. *Environmental Toxicology and Chemistry* 4:209–217.

Bernardi, G., P. Sordino, and D. A. Powers. 1993. Concordant mitochondrial and nuclear DNA phylogenies for populations of the teleost fish *Fundulus heteroclitus*. *Proceedings of the National Academy of Sciences* 90:9271–9274.

Bradley, R. W., C. DuQuesnay, and J. B. Sprague. 1985. Acclimation of rainbow trout, *Salmo gairdneri* Richardson, to zinc: Kinetics and mechanism of enhanced tolerance induction. *Journal of Fish Biology* 27:367–379.

Brown, D. C., I. J. Ropson, and D. A. Powers. 1988. Biochemical genetics of *Fundulus heteroclitus* (L.). V. Inheritance of 10 biochemical loci. *Journal of Heredity* 79:359–365.

Brummet, A. R. 1966. Observations on the eggs and breeding season of *Fundulus heteroclitus* at Beaufort, North Carolina. *Copeia* 1966:616–620.

Brummett, A. R., and J. N. Dumont. 1981. A comparison of chorions from eggs of northern and southern populations of *Fundulus heteroclitus*. *Copeia* 1981:607–614.

Bush, C. P., and J. S. Weis. 1983. Effects of salinity on fertilization success in two populations of *Fundulus heteroclitus*. *Biological Bulletin* 164:406–417.

Diamond, S. A., M. C. Newman, M. Mulvey, and S. I. Guttman. 1991. Allozyme genotype and time-to-death of mosquitofish, *Gambusia holbrooki*, during acute inorganic mercury exposure: a comparison of populations. *Aquatic Toxiocology* 21:119–134.

DiMichele, L., and D. A. Powers. 1984. Developmental and oxygen consumption rate differences between lactate dehydrogenase-B genotypes of *Fundulus heteroclitus* and their effect on hatching time. *Physiological Zoology* 57:52–56.

Donker, M. H., C. Zonneveld, and N. M. van Straalen. 1993. Early reproduction and increased reproductive allocation in metal-adapted populations of the terrestrial isopod *Porcellio scaber. Oecologia* 96:316–323.

Forbes, V. E. 1996. Chemical stress and genetic variability in invertebrate populations. *Toxicology and Ecotoxicology News* 3:136–141.

Foster, P. L. 1977. Copper exclusion as a mechanism of heavy metal tolerance in a green alga. *Nature* 269:322–323.

Gonzales-Villasenor, L. I., and D. A. Powers. 1990. Mitochondrial DNA restriction site polymorphism in the teleost *Fundulus heteroclitus* support secondary intergradation. *Evolution* 44:27–37.

Heagler, M. G., M. C. Newman, M. Mulvey, and P. M. Dixon. 1993. Allozyme genotype in mosquitofish, *Gambusia holbrooki*, during mercury exposure: temporal stability, concentration effects, and field verification. *Environmental Toxicology and Chemistry* 12:385–395.

Hebert, P. D., M. M. Luiker. 1996. Genetic effects of contaminant exposure—towards an assessment of impacts on animal populations. *Science of the Total Environment* 191:23–58.

Hoare, K., A. R. Beaumont, and J. Davenport. 1995. Variation among populations in the resistance of *Mytilus edulis* embryos to copper: adaptation to pollution? *Marine Ecology Progress Series* 120: 155–161.

Khan, A. T., and J. S. Weis. 1987a. Effects of methylmercury on sperm and egg viability of two populations of killifish, *Fundulus heteroclitus*. *Archives of Environmental Contamination and Toxicology* 16:499–505.

Khan, A. T., and J. S. Weis. 1987b. Effects of methylmercury on egg and juvenile viability in two populations of the killifish, *Fundulus heteroclitus*. *Environmental Research* 44:272–278.

Khan, A. T., and J. S. Weis. 1993a. Bioaccumulation of heavy metals in two populations of mummichog (*Fundulus heteroclitus*). *Bulletin of Environmental Contamination and Toxicology* 51:1–5.

Khan, A. T., and J. S. Weis. 1993b. Differential effects of organic and inorganic mercury on the micropyle of the eggs of *Fundulus heteroclitus*. *Environmental Biology of Fishes* 37:323–327.

Klerks, P. L., and J. S. Weis. 1987. Genetic adaptation to heavy metals in aquatic organisms. *Environmental Pollution* 45:173–205.

Lavie, B., and E. Nevo. 1986. Genetic selection of homozygote allozyme genotypes in marine gastropods exposed to cadmium pollution. *Science of the Total Environment* 57:91–98.

Leary, R. F., F. W. Allendorf, and K. L. Knudsen. 1985. Inheritance of meristic variation and the evolution of developmental stability in rainbow trout. *Evolution* 39:308–314.

Lee, C. L., M. C. Newman, and M. Mulvey. 1992. Time to death of mosquitofish (*Gambusia holbrooki*) during acute inorganic mercury exposure: Population structure effects. *Archives of Environmental Contamination and Toxicology* 22:284–287.

Longwell, A. C., and J. B. Hughes. 1980. Cytologic, cytogenetic, and developmental state of Atlantic mackerel eggs from sea surface waters of the New York Bight, and prospects for biological effects monitoring with ichthyoplankton. *Rapports et Proces-Verbaux des Reunions Conseil International pour l' Exploration de la Mer* 179:275–291.

Magnusson, J., and C. Ramel. 1986. Genetic variation in the susceptibility to mercury and other metal compounds in *Drosophila melanogaster. Teratogenesis Carcinogenesis and Mutagenesis* 6:289–305.

McFarlane, G. A, and W. G. Franzin. 1978. Elevated heavy metals: a stress on a population of white suckers, *Catostomus commersoni*, in Hamell Lake, Saskatchewan. *Journal of the Fisheries Research Board of Canada* 35:963–970.

Mitton, J. B., and R. K. Koehn. 1975. Morphological adaptation to thermal stress in a marine fish, *Fundulus heteroclitus*. *Biological Bulletin* 151:548–559.

Moraitou-Apostolopoulou, M. M., G. Verriopoulos, and I. Rogdakis. 1982. Evaluation of the stress exerted by a polluted environment to a marine organism by comparative toxicity tests. *Bulletin of Environmental Contamination and Toxicology* 28:416–423.

Morin, R. P., and K. W. Able. 1983. Geographic variation in the egg morphology of the cyprinodontid fish, *Fundulus heteroclitus*. *Copeia* 1983:726–740.

Mugue, N. 1996. Population genetics of *Fundulus heteroclitus* (L.) in New York/New Jersey Harbor estuary. Ph.D. Dissertation. Newark, NJ: Rutgers University.

Mulvey, M., M. C. Newman, A. Chazal, M. M. Keklak, M. G. Heagler, and S. H. Hales. 1995. Genetic and demographic responses to mosquitofish (*Gambusia holbrooki* Girard 1859) populations stressed by mercury. *Environmental Toxicology and Chemistry* 14:1411–1418.

Munkittrick, K. R., and D. G. Dixon. 1988. Evidence for a maternal yolk factor associated with increased tolerance and resistance to feral white sucker (*Catostomus commersoni*) to waterborne copper. *Ecotoxicology and Environmental Safety* 15:7–20.

Munzinger, A., and F. Monicelli. 1992. Heavy metal co-tolerance in a chromium tolerant strain of *Daphnia magna*. *Aquatic Toxicology* 23:203–216.

Nevo, E., T. Shimony, and M. Libni. 1978. Pollution selection of allozyme polymorphisms in barnacles. *Experientia* 34:1562–1564.

Nevo, E., T. Perl, and D. Wool. 1981. Mercury selection of allozyme genotypes in shrimps. *Experientia* 37:1152–1154.

Nevo, E., R. Ben-Shlomo, and B. Lavie. 1984. Mercury selection of allozymes in marine organisms. *Proceedings of the National Academy of Science* 81:1258–1259.

Newman, M. C., S. A. Diamond, M. Mulvey, and P. Dixon. 1989. Allozyme genotype and time to death of mosquitofish, *Gambusia affinis* (Baird and Girard) during acute toxicant exposure: A comparison of arsenate and inorganic mercury. *Aquatic Toxicology* 15:141–156.

Newman, M. C., M. Mulvey, A. Beeby, R. W. Hurst, and L. Richmond. 1994. Snail (*Helix aspersa*) exposure history and possible adaptation to lead as reflected in shell composition. *Archives of Environmental Contamination and Toxicology* 27:346–351.

Pascoe, D., and J. H. Beattie. 1979. Resistance to cadmium by pre-treated rainbow trout alevins. *Journal of Fish Biology* 14:303–306.

Perry D., J. S. Weis, and P. Weis. 1988. Cytogenetic effects of methylmercury in embryos of the killifish, *Fundulus heteroclitus*. *Archives of Environmental Contamination and Toxicology* 17:569–574.

Pierce, B. A., and D. K. Wooten. 1992. Genetic variation in tolerance of amphibians to low pH. *Journal of Herpetology* 26:422–429.

Place, A. R., and D. A. Powers. 1978. Genetic bases for protein polymorphysm in *Fundulus heteroclitus* (L.). I. Lactate Dehydrogenase (*Ldh*-B), Malate Dehydrogenase (*Mdh*-A), Glucosephosphate Isomerase (*Gpi*-B), and Phosphoglucomutase (*Pgm*-A). *Biochemical Genetics* 16:578–591.

Powers, D. A., and A. R. Place. 1978. Biochemical genetics of *Fundulus heteroclitus* (L.). I. Temporal and spatial variation in gene frequencies of *Ldh*-B, *Mdh*-A, *Gpi*-B, and *Pgm*-A. *Biochemical Genetics* 16:593–607.

Powers, D. A., I. Ropson, D. Brown, R. Van Beneden, R. Cashon, L. I. Gonzalez-Villasenor, and L. A. DiMichele. 1986. Genetic Variation in *Fundulus heteroclitus*: Geographic distribution. *American Zoologist* 26:131–144.

Prince, R., and K. R. Cooper. 1995. Comparisons of the effects of 2,3,7,8-tetrachlorodibenzo-*p*-dioxin on chemically impacted and nonimpacted subpopulations of *Fundulus heteroclitus*: 1. TCDD toxicity. *Environmental Toxicology and Chemistry* 14:579–587.

Prinslow. T. E., I. Valiela, and J. E. Teal. 1974. The effect of detritus and ration size on the growth of *Fundulus heteroclitus* (L). *Journal of Experimental Marine Biology and Ecology* 16:1–10.

Rahel, F. J. 1981. Selection for zinc tolerance in fish: Results from laboratory and wild populations. *Transactions of the American Fisheries Society* 110:19–28.

Ropson, I. J., D. C. Brown, and D. A. Powers. 1990. Biochemical genetics of *Fundulus heteroclitus* (L.). VI. Geographical variation in the gene frequencies of 15 loci. *Evolution* 44:16–26.

Smith, G. M., A. T. Khan, J. S. Weis, and P. Weis. 1995. Behavior and brain chemistry correlates in mummichogs (*Fundulus heteroclitus*) from polluted and unpolluted environments. *Marine Environmental Research* 39:329–333.

Smith, G. M., and J. S. Weis. 1997. Predator/prey interactions in *Fundulus heteroclitus*: Effects of living in a polluted environment. *Journal of Experimental Marine Biology and Ecology* 209:75–87.

Smith, P. J., and Y. Fujio. 1982. Genetic variation in marine teleosts: High variability in habitat specialists and low variability in habitat generalists. *Marine Biology* 69:7–20.

Stebbing, A. R. D. 1981. Hormesis—stimulation of colony growth in *Campanularia flexuosa* (Hydrozoa) by copper, cadmium, and other toxicants. *Aquatic Toxicology* 1:227–238.

Stuhlbacher, A., M. C. Bradley, C. Naylor, and P. Calow. 1992. Induction of cadmium tolerance in two clones of *Daphnia magna* Straus. *Comparative Biochemistry and Physiology C Comparative Pharmacology and Toxicology* 101:571–577.

Swofford, D. L., and R. K. Selander. 1981. BIOSYS-1: a Fortran program for the comprehensive analysis of electrophoretic data in population genetics and systematics. *Journal of Heredity* 72:281–283.

Tave, D. 1984. Genetics of dorsal fin ray number in the guppy, *Poecilia reticulata*. *Copeia* 1984:140–144.

Toppin, S. V., M. Heber, J. S. Weis, and P. Weis. 1987. Changes in reproductive biology and life history in *Fundulus heteroclitus* in a polluted environment. In *Pollution physiology of estuarine organisms*, eds. W. Vernberg, A. Calabrese, F. Thurberg, and F. J. Vernberg, 171–184. Columbia: University of South Carolina Press.

Tranvik, L., G. Bengtsson, and S. Rundgren. 1993. Relative abundance and resistance traits of two Collembola species under metal stress. *Journal of Applied Ecology* 30:43–52.

van Veld, P. A., and D. J. Westbrook. 1995. Evidence for depression of cytochrome P450A in a population of chemically resistant mummichog (*Fundulus heteroclitus*). *Environmental Science* 3,4:221–234.

Weis, J. S., and A. A. Khan. 1990. Effects of mercury on the feeding behavior of the mummichog, *Fundulus heteroclitus* from a polluted habitat. *Marine Environmental Research* 30:243–249.

Weis, J. S., and A. A. Khan. 1991. Reduction in prey capture ability and condition of mummichogs from a polluted habitat. *Transactions of the American Fisheries Society* 120:127–129.

Weis, J. S., M. Renna, S. Vaidya, and P. Weis. 1987. Mercury tolerance in killifish: a stage-specific phenomonon. In *Oceanic processes in marine pollution Vol. 1. Biological processes and wastes in the ocean*, eds. J. Capuzzo and D. Kester, 31–36. Malabar, FL: Krieger Publishing Co.

Weis, J. S., P. Weis, M. Heber, and S. Vaidya. 1981a. Methylmercury tolerance of killifish (*Fundulus heteroclitus*) embryos from a polluted vs non-polluted environment. *Marine Biology* 65:283–287.

Weis, J. S., P. Weis, M. Heber, and S. Vaidya. 1981b. Investigations into mechanisms of methylmercury tolerance in killifish (*Fundulus heteroclitus*) embryos. In *Physiological mechanisms of marine pollutant toxicity*, eds. W. Vernberg, A. Calabiese, F. P. Thurberg, and F. J. Vernberg, 311–330. New York: Academic Press.

Weis, J. S., P. Weis, and J. L. Ricci. 1981c. Effects of cadmium, zinc, salinity and temperature on the teratogenicity of methylmercury to the killifish (*Fundulus heteroclitus*). *Rapports et Proces-Verbaux des Reunions Conseil International pour l'Exploration de la Mer* 178:64–70.

Weis, J. S., P. Weis, and M. Heber. 1982a. Variation in response to methylmercury by killifish, (*Fundulus heteroclitus*) embryos. In *Aquatic toxicology and hazard assessment: fifth conference*, eds. J. G. Pearson, R. Foster, and W. E. Bishop, 109–119. ASTM STP 766, Philadelphia, PA: American Society for Testing and Materials.

Weis, J. S., and P. Weis. 1984. A rapid change in methylmercury tolerance in a population of killifish, *Fundulus heteroclitus*, from a golf course pond. *Marine Environmental Research* 13:231–245.

Weis, J. S., P. Weis, M. Renna, and S. Vaidya. 1985. Search for a physiological component of methylmercury tolerance in the mummichog, *Fundulus heteroclitus*. In *Marine pollution and physiology: recent advances*, eds. F. J. Vernberg, F. P. Thurberg, A. Calabrese, and W. B. Vernberg, 309–326. Columbia: University of South Carolina Press.

Weis, P. 1984. Metallothionein and mercury tolerance in the killifish, *Fundulus heteroclitus*. *Marine Environmental Research* 14:153–166.

Weis, P., and J. S. Weis. 1977. Methylmercury teratogenesis in the killlifish, *Fundulus heteroclitus*. *Teratology* 16:317–326.

Weis, P., and J. S. Weis. 1978. Methylmercury inhibition of fin regeneration in fishes and its interaction with salinity and cadmuim. *Estuarine and Coastal Marine Science* 6:327–334.

Weis, P., and J. S. Weis. 1982a. Toxicity of methylmercury, mercuric chloride and lead in killifish (*Fundulus heteroclitus*) from Southampton, New York. *Environmental Research* 28:364–374.

Weis, P., and J. S. Weis. 1982b. Toxicity of the PCBs Aroclor 1254 and 1242 to embryos and larvae of the mummichog, *Fundulus heteroclitus*. *Bulletin of Environmental Contamination and Toxicology* 28:298–304.

Weis, P., and J. S. Weis. 1983. Effects of embryonic pre-exposure to methylmercury, $Hg^{2+}$ and $Cu^{2+}$ on larval tolerance in *Fundulus heteroclitus*. *Bulletin of Environmental Contamination and Toxicology* 31:530–534.

Zhou, T. 1997. Behavioral development and neurobehavioral effects of methylmercury on larval mummichogs, *Fundulus heteroclitus*, from polluted and reference environments. Ph.D. dissertation, Newark, NJ: Rutgers University.

*Genetics and Ecotoxicology*
Edited by V. E. Forbes
Copyright © 1999 Taylor & Francis

# 4

# Chemical-Induced Changes in the Genetic Structure of Populations: Effects on Allozymes

## Robert B. Gillespie and Sheldon I. Guttman

**Abstract.** Drawing from evidence on research in aquatic toxicology, we propose that allozyme analysis can be a useful tool for estimating the effects of environmental chemicals on genetic variation in natural populations. Recent research has associated exposure to environmental toxicants with changes in allozyme variation in populations of many species. Additionally, laboratory evidence has shown that variation in allozyme genotypes may either be adaptive to exposure to environmental toxicants or may be a marker for changes in genetic variation in response to environmental pollutants. Evidence is also presented to suggest that changes in allozyme variation are associated with significant changes in heterozygosity and polymorphism and that these changes may indicate an increase in susceptibility to negative impacts from further environmental perturbations. Because results from research studies are not equivocal, more effort is needed to test the ability of allozyme variation to reflect contaminant-induced changes in genetic variation in natural populations. However, the evidence presented here and elsewhere suggests that allozyme analysis may already be useful in estimating the impacts of environmental contaminants on genetic variation in natural populations.

**Keywords.** Adaptation; allozyme frequency; genetic polymorphism; heterozygosity.

## INTRODUCTION

Since the early 1900s, research has shown that environmental toxicants significantly change genetic variation in natural populations (Morton 1993). Although much of this evidence comes from research on the evolution of resistance to insecticides, recent studies on many species have associated exposure to other environmental contaminants with changes in allelic variation of enzymes (allozymes). Although the mechanisms are difficult to ascertain from field studies, evidence suggests that changes in allozyme variation could come either from *bottleneck effects* or from *selection effects* (Murdoch and Hebert 1994).

Laboratory studies have shown that organisms with different allozyme genotypes vary in their tolerance to the toxicity of various toxicants (Gillespie and Guttman 1993a; Guttman 1994). Therefore, in addition to changes in allozyme variation that are due to severe reductions in effective population size, specific allozyme genotypes may be selected for because they contribute to resistance to the toxic effects of pollutants. Alternatively, if allozyme variation is not directly adaptive or contributes in a minor way to genetically-based resistance, then allozyme analysis could still be used as a marker of genetic susceptibility or as an estimate of genetic variation in populations. Because genetic variation is an ecologically important variable, analysis of allozyme variation may be a useful parameter to assess the impacts of environmental contaminants on ecosystems.

Changes in allozyme frequencies may not necessarily result in negative effects for populations. In fact, if survival of the population under environmental stress is dependent upon specific resistant allozyme genotypes, then changes in allozyme frequencies are adaptive. However, if changes in allozyme frequencies in the surviving population are concomitant with a reduction in overall genetic variation (e.g., less heterozygosity and polymorphism) or increased susceptibility to further environmental stressors, then these changes could be associated with negative effects (Guttman 1994; Gillespie 1996).

Several researchers have proposed that allozyme variation be monitored in populations of organisms that are susceptible to environmental chemical pollutants (Evenden and Depledge 1997; Gillespie and Guttman 1993a; Ben-Shlomo and Nevo 1988; Gillespie 1996; Kopp et al. 1994). If allozyme variation is a reliable measure of overall genetic variation, then monitoring this parameter in natural populations could be valuable for risk analysis in ecotoxicology (Evenden and Depledge 1997; Guttman 1994).

The objective of this chapter is to provide evidence that environmental chemical contaminants are associated with allozyme variation in natural populations. Furthermore, we propose that changes in allozyme frequencies may be used as an indirect measure of loss of genetic variation and that changes in frequencies themselves may be associated with negative effects in natural populations. The evidence we present in this chapter was organized to establish three points about the relationship between environmental toxicants and allozyme variation:

1. Changes or differences in allozyme frequencies in natural populations are correlated with exposure to chemical contaminants.
2. Individuals with different allozyme genotypes are more susceptible to the toxicity of contaminants than those with other genotypes.
3. Changes in allozyme frequencies in surviving populations are associated either with a loss of genetic variation or with an increased susceptibility to further environmental stressors.

## CHEMICAL TOXICITY AND EFFECTS ON GENETIC VARIATION

There are at least two major processes whereby chemical toxicants can cause significant changes in genetic variation in natural populations (Fig. 1). Exposure of a population to toxic pollutants may result in a significant reduction in effective population

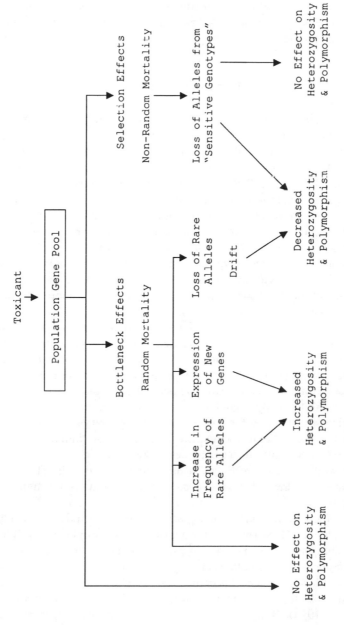

FIG. 1. Conceptual representation of possible effects of long-term exposure to environmental toxicants on genetic variation in a population.

size causing a *bottleneck effect*. Alternatively, differential mortality from toxicity could cause *selection effects* that would favor "tolerant genotypes" and tend to eliminate "sensitive genotypes."

## Bottleneck Effects

Drastic reductions in effective population size should result in a decrease in genetic variation due to the loss of low frequency and rare alleles (Lande 1988). Additionally, it would be expected that genetic drift would tend to fix some alleles and eliminate others. Evidence from field studies has documented a loss of genetic variation for certain animals such as cheetahs (O'Brien et al. 1987). However, other evidence suggests that bottleneck effects can cause an increase in variation of quantitative characteristics that may stem from an increase in genetic variation (Lewin 1987). Inbreeding in a small population could increase the proportion of previously rare alleles and make them more common. The contribution of new alleles and new epistatic interactions among genes could increase phenotype variation of quantitative traits in a bottlenecked population. However, previously rare alleles may also increase the proportion of maladaptive allele combinations in the population, leaving it more susceptible to other environmental stressors (Lewin 1987).

Acute exposure to toxic contaminants could cause significant mortality that would drastically reduce a population's effective size. However, distinguishing bottleneck effects from other effects in the field is difficult. One way to detect contaminant-induced bottleneck effects would be to document that genetic variation in the impacted population differs from a reference or pre-exposure population by the loss of low-frequency and rare alleles. Additionally, a bottleneck effect could be inferred if a decrease in genetic variation was detected for several impacted populations exposed to similar contaminants. If allele and genotype frequencies among these similarly-exposed populations differed significantly, then a bottleneck may be implicated because selection for resistance would tend to create genetically similar individuals. This possibility "boldly" assumes that the initial pre-exposure genetic variation was similar among distinct populations.

Bottleneck effects from contaminant exposure have been implicated in only a few studies. Strittholt et al. (1988) proposed that a lack of polymorphism in yellow perch of Lake Erie could have resulted from a contaminant-induced bottleneck effect. Similarly, a contaminant-induced bottleneck may have contributed to a loss of genetic variation in the Florida panther (Facemire et al. 1995). Acute toxicity to rotenone significantly reduced the effective population size of river chubs in the Virgin River (DeMarais et al. 1993). Because the changes in allozyme genotypes between pre-exposed and post-exposed populations of chubs differed among sites, the authors proposed that a bottleneck effect, rather than selection, caused changes in allozyme variation. Similar arguments were made to suggest a bottleneck effect for significantly less haplotype variation of mitochondrial DNA in brown bullheads sampled from polluted tributaries of the Great Lakes (Murdoch and Hebert 1994). Although a bottleneck effect was implicated in the studies above, the evidence is mostly circumstantial rather than directly empirical.

Theoretically, a contaminant-induced bottleneck event should decrease allozyme variation in natural populations. However, a novel environment created by exposure to toxicants could cause expression of new genes that could increase the phenotype variation of fitness characteristics, such as growth rate and development (Holloway et al. 1990). Thus, at least in the short-term, genetic variation may increase. The evidence presented above suggests that the outcome of a contaminant-induced bottleneck on genetic variation is not easily predicted, especially in natural populations where other factors can affect allele frequencies. It is also possible that bottleneck effects and selection effects could act together to affect genetic variation in an exposed population that would make distinctions of effects complicated. The effect of contaminant-induced bottlenecks on genetic variation in populations is in need of further research.

### Genetic Basis for Tolerance to Toxicants

The genetic basis for resistance to chemical toxicants has been studied extensively for mechanisms that confer resistance to pesticides in insects. From these studies we know that resistance to toxicants has been attributed to genetic variation in targets of toxicity and to variation in toxicity-modifying factors (Morton 1993; Kasai and Hanazato 1995). Target molecules of toxicants include cell receptors, acetylcholinesterase, sodium channels, metabolic enzymes, hormones, and nucleic acids (Morton 1993). Modifiers of toxicity include cell membrane structure, stress proteins, mixed-function oxidase enzymes, esterases, and glutathione S-transferase. Allelic variation at any of the loci that contribute to phenotypic variation in toxicity targets or in toxicity modification factors could contribute to the pool of variation that can be acted upon by selection.

Resistance to toxicants could evolve from allelic variation at single loci or at multiple loci depending upon the exposure regime. If so, the effect of toxicant-induced selection on genetic variation may depend upon the genetic basis for resistance. In some cases, the genetic basis for resistance to the toxic effects of chemicals results from small variation at one or two genes (Mouches et al. 1986; ffrench-Constant et al. 1993; Schat et al. 1993). If survivors with resistant alleles vary at other loci, then overall genetic variation (e.g., heterozygosity and polymorphism) in an impacted population may not be affected (Fig. 1). On the other hand, if resistance to toxicants is highly polygenic, then contaminant-induced selection could decrease genetic variation. Depending upon the genetic basis for resistance, selection for tolerance to environmental toxicants in natural populations could lead to a range of effects from no change in genetic variation to very significant changes in genetic variation.

Heritability of toxicity resistance to environmental pollutants has been observed in natural populations of several species and different toxicants (Mulvey and Diamond

1991). For example, oligochaetes evolved resistance to heavy metals in sediments over a short period of time (Klerks and Levinton 1989). Resistance to metal toxicity was attributed, in part, to the ability of resistant worms to sequester metals in the cytosol. Although resistance was inherited, second generation worms lost some resistance. Thus, if the loss of resistance resulted from selection relaxation, then resistance adaptation may come with fitness costs (Klerks and Levinton 1989).

Evolution for toxicant resistance in natural populations is not universal and the contribution of environmental factors to variation in toxicity resistance cannot be ruled out (Pierce and Wooten 1992; Klerks and Levinton 1989). Although the relative contribution of genes and environment to resistance may be unclear, even low levels of heritability could lead to evolution of resistance in organisms (Stuhlbacher et al. 1993; Baird et al. 1991, 1990). However, the rate of selection for resistance and the significance of its effects on genetic variation in the population may depend upon the degree of inheritance and the number of genes contributing to the phenotype of resistance (Mulvey and Diamond 1991).

### Effects of Selection on Genetic Variation

The effects of selection for toxicity tolerance on genetic variation should be dependent upon the type of toxicant, strength of selection (concentration), and duration of exposure. Initial selection for resistance will depend upon existing additive genetic variation of resistance mechanisms. Thus, in the short-term, impacted populations may differ from unexposed populations by only a few genes. However, over long-term exposure, new alleles from mutations and immigration can contribute to genetic variation that could allow for further resistance adaptation (Morton 1993). Thus, after prolonged exposure to contaminants, impacted populations could differ from unexposed populations by several independent genes. The strength of selection will also have an impact on the effects of resistance adaptation to genetic variation. Acute exposure (strong selection force) to toxicants should cause rapid resistance adaptation by selection at a single or a few genes that have "large effects" (Morton 1993). However, long-term sublethal exposure (weak selection force) may cause slow resistance evolution at polygenic loci. The effects of contaminant-induced selection on genetic variation may depend upon the number of loci contributing to resistance and the strength and duration of the selection force.

It is possible, in the short-term, that low-level concentrations of toxicants may initially increase genetic variation in a population. Short-term increases in genetic variation could arise from an initial increase in the proportion of previously rare but tolerant alleles in the population. Over extended periods of time, however, strong contaminant-induced selection for resistant genotypes should decrease genetic variation in a population by eliminating alleles carried by sensitive genotypes. Evidence for a selection-induced decrease in genetic variation could come from estimates of heterozygosity and polymorphism using allozyme analysis.

## Significance of Effects

Contaminant-induced changes in genetic variation in a population may significantly increase the risk of its extinction by lowering its ability to adapt (Fig. 1). Susceptibility to extirpation could come from either a significant loss of genetic variation or increased susceptibility to environmental stress due to selection for resistance (Evenden and Depledge 1997). A loss of alleles from mortality of "sensitive" individuals, could reduce the adaptive potential of a population by eliminating genotypes that could survive novel environmental challenges or help the population recover after the chemical stressor is remediated.

A loss of "sensitive" alleles may reduce the fitness of the population. Energy allocation between survival (resistance) and growth may cause an energetic trade-off between tolerance to toxicity and reproduction in animals (Holloway et al. 1990). Thus, in uncontaminated environments, we would expect "sensitive" but fast-growing individuals to contribute to the population. But in a toxicant-stressed environment, we would expect a larger frequency of slower growing but resistant individuals. Therefore, although chemical induced selection might cause evolution for resistant geno types that allows the population to survive, the loss of "sensitive" alleles may reduce the adaptive potential of the population.

A significant loss of genetic variation (heterozygosity and polymorphism) and increased susceptibility to environmental stressors (decreased fitness) are negative effects that can occur without obvious detection. Analyzing allozyme variation in populations may be a valuable method for estimating genetic variation in a population. Therefore, monitoring the allozyme variation in populations may help detect potential negative effects before more serious consequences occur.

## Allozymes, Environmental Toxicants, and Genetic Variation

Allozyme variation has been used for over 30 years to quantify genetic variation in populations of many species of organisms (Avise 1994). Electrophoretic detection of allozyme analysis allows rapid quantification of genetic variation that is only two steps away from the gene (Table 1). Variation at many enzyme-coding loci has been discovered for many species, and techniques for allozyme analysis are easily learned (Gillespie 1996).

The value of allozyme variation in studying evolution has been controversial. Although many species exhibit allelic variation at several enzyme loci, the adaptive value of this variation is a point of debate. Observed differences in allozyme frequencies in nature could result from stochastic processes, drift, gene flow, selection, or any combination of these processes. The neutralist theory proposes that allozyme variation is not selectively adaptive and mathematical models based on neutrality can account for protein variation in nature (Kimura 1993). Additionally, it has been argued that the variation in activity of allozymes, as detected by electrophoretic mobility, may be small enough to make no difference in enzyme flux and thus no

**TABLE 1.** *Diagrammatic representation of parameters that could be used to measure genetic variation in organisms*

| Relative genetic significance | Very high | | | Very low |
|---|---|---|---|---|
| Genetic parameter | Gene sequence variation | nDNA, mtDNA structure variation | Protein variation | Morphological character variation |
| Analytical technique | Nucleotide sequences | Restriction fragment length polymorphism | Allozyme analysis | Meristic characters |
| | | Random amplified polymorphic DNA | Amino acid sequences | Color and morphometry |
| Relative functional significance | Very low | | | Very high |

difference in fitness of individuals with different allozyme genotypes (Eanes 1987). Others have argued that differences in the function of allozymes are linked with differences in fitness (Watt 1994). Evidence for the adaptive potential of allozyme variation comes from studies that linked differences in thermal stability of allozyme variants with variation of allozyme frequencies of populations in different thermal habitats (Mitton and Koehn 1975; Nevo et al. 1977; Watt 1977; Graves and Somero 1982; Smith et al. 1983; van Beneden and Powers 1989). A selective potential for thermal stability does not necessarily guarantee that allozyme variation is selectively adaptive for contaminant-induced effects. However, evidence discussed later suggests that allozyme variation may contribute to selective adaptation to tolerance to toxicants although the mechanisms for adaptive value of allozymes are still a matter of study. Even if allozyme variation itself is not directly adaptive, allozyme variation may be a valuable genetic marker for susceptibility to toxicants and may be a useful estimate of genetic variation in populations (Guttman 1994).

The subsequent discussion presents the results of research on the relationship between exposure to environmental toxicants and allozyme variation. The body of evidence we present is not meant to be an exhaustive review of the literature, but rather a sample of studies that support the utility of allozyme analysis in the study of the effects of environmental pollutants on genetic variation. Reviews on the effects of pesticides on genetic variation in insects and on the development of toxicant resistance in plants have been presented elsewhere (for plants see Shaw, chapter 2). Because we have not encountered a comprehensive review of this literature, we concentrated on research that studied the relationship between environmental contaminants and allozyme variation in aquatic animals. However, we believe that the evidence and arguments presented with these data can be applied to any environmentally contaminated system.

## EVIDENCE FOR EFFECTS OF CHEMICALS
## ON ALLOZYME FREQUENCIES

The evidence that chemical contaminants affect allozyme frequencies in natural populations comes from field and laboratory studies (Guttman 1994). Field studies have provided indirect evidence that contaminant toxicity can result in selection- or bottleneck-induced changes in allozyme frequencies. However, because these studies are primarily *ecological epidemiology*, they cannot provide cause-effect relationships. Laboratory studies, on the other hand, have produced more controlled evidence that exposure to toxicants can cause changes in allozyme frequencies through variable survival of individuals with different allozyme genotypes.

### Allozyme Frequencies and Chemical Exposure in Natural Populations

Many studies conducted over the past 30 years have shown associations between allozyme variation and exposure to chemical contaminants (Table 2). As with any field study, it is difficult to show that chemical toxicants are the sole cause of changes in allozyme frequencies. Other factors such as drift and gene flow dynamics can cause differences in allozyme frequencies (Avise 1994). However, unless evidence can be detected in natural populations, there is little support for the hypothesis that environmental toxicants can significantly affect genetic variation in nature. It is difficult to draw obvious generalizations about the effect of contaminants on allozyme variation in natural populations because the local conditions, species studied, and study designs vary. However, an examination of this list of studies suggests that allozyme variation in a variety of taxa has been shown to vary with exposure to environmental chemicals. Additionally, associations between environmental contaminants and allozyme frequencies have been shown for a variety of chemical pollutants.

Perhaps the most interesting information that comes from these field studies is the apparent similarity of the enzyme loci that vary in allele frequencies with environmental contaminant exposure. Of the studies presented here, the enzyme loci most often associated with correlations between chemical toxicants and allozyme frequency changes include phosphoglucomutase (*Pgm*), glucose phosphate isomerase (*Gpi*), and malate dehydrogenase (*Mdh*). It is tempting to conclude from these studies that allelic variation at these enzyme loci is associated with chemical-induced selection for tolerant genotypes. However, these loci may simply be the most common loci analyzed because of their relatively high variability in many species. Alternatively, these loci may be linked with others that are associated with genetically-based tolerance to chemical toxicity and are therefore reliable markers of effects.

Common enzyme loci have been associated with exposure to chemical toxicants in many species and for many different types of pollutants. Despite potential confounding factors, the possibility that specific enzyme loci may be consistently associated with environmental toxicity and changes in allozyme frequencies justifies closer examination.

**TABLE 2.** *A list of species and chemical toxicants that were associated with differences or changes in allozyme frequencies in natural populations*

| Chemical | Taxon | Species | Enzyme | Reference |
|---|---|---|---|---|
| Mercury | Shrimp | *Palaemon elegans* | *Pgm* | Nevo et al. 1984 |
| | Gastropods | *Monodonta turbinata* | *Gpi* | Nevo et al. 1984 |
| | Mosquitofish | *Gambusia affinis* | *Gpi* | Heagler et al. 1993 |
| Uranium | Mosquitofish | *Gambusia affinis* | *Idh* | Keklak et al. 1994 |
| | | | *Gpi* | |
| | | | *Mpi* | |
| Heavy metals | Freshwater snail | *Heliosoma trivolvis* | *Gpi* | Benton et al. 1994 |
| | Mosquitofish | *Gambusia holbrooki* | *Gpi* | Benton et al. 1994 |
| | | | *Idh* | |
| | | | *Acp* | |
| | | | *Me* | |
| | Soil arthropod[1] | *Orchesella bifasciata* | *Pgm* | Tranvik et al. 1994 |
| | | | *Gpi* | |
| Complex effluent | Stoneroller minnow | *Campostoma anomalum* | *Pgm* | Gillespie & Guttman 1989 |
| | | | *Mdh* | |
| Complex effluent | Spotfin shiner | *Notropis spilopterus* | *Gpi* | Gillespie & Guttman 1993b |
| Water quality | Bluntnose minnow | *Pimephales notatus* | *Est*[2] | Fore et al. 1995a |
| | | | *Gpi* | |
| | Stoneroller minnow | *Campostoma anomalum* | *Ada* | Fore et al. 1995b |
| | | | *Est* | |
| | | | *Pgm* | |
| | | | *Mdh* | |
| | | | *Gpi* | |
| | | | *Tpi* | |
| | | | *Acp* | |
| Acidic waters | Central mudminnow | *Umbra limi* | *Idh* | Kopp et al. 1992 |
| | | | *Pgd* | |
| | | | *Mdh* | |
| | | | *Gpi* | |
| | | | *Mpi* | |
| Rotenone | River chub | *Gila seminuda* | *Cbp* | DeMarais et al. 1993 |
| | | | *Ck* | |
| | | | *Pgm* | |
| Insecticides | Mosquitofish | *Gambusia affinis* | *Gpi* | Hughes et al. 1991 |

[1] Only minor differences were found in allozyme frequencies between contaminated and reference populations.

[2] Differences in allozymes that were associated with water quality included the presence/absence of the *Est*-3 locus.

Because field studies can only produce correlations between allozyme frequencies and exposure to environmental pollutants, the hypothesis that chemical toxicants cause significant changes in allozyme variation is only partly supported. Controlled studies that show evidence for differential tolerance of allozyme genotypes to toxicity from chemical toxicants are necessary to establish a cause-effect relationship.

### Allozyme Genotypes and Differential Resistance to Toxicants

Laboratory studies have associated differential sensitivity to toxicants with allozyme genotypes (Table 3). A variety of contaminants have been associated with variable resistance to toxicity for different allozyme genotypes in many species. However, studies on the relationship between allozyme genotype and differential sensitivity to metals dominate the literature.

Because metals can inhibit certain enzymes (Milstein 1961a,b; Mizrahi and Achituv 1989), it is possible that allozyme variants of enzymes differ in their ability to resist inhibition by metals. If so, then allozyme variation could be adaptive to a population exposed to metal pollution. Chagnon and Guttman (1989a) found that in vitro enzyme inhibition thresholds to copper by *Pgm* exposed to copper varied among different fish species. Additionally, differentially-inhibited isoalleles of *Pgm* were detected for copper in mosquitofish. Variation in inhibition thresholds among species and between isoalleles for *Pgm* was hypothesized to result from differential resistance of allozymes to inhibition by copper (Chagnon and Guttman 1989a). Furthermore, the authors suggested that differential enzyme inhibition may be responsible for variation in sensitivity to toxicity of metals by allozyme genotypes of mosquitofish (Chagnon and Guttman 1989b).

We have not found studies that provide direct evidence for the differential enzyme inhibition hypothesis (Table 4). Furthermore, two studies showed that a reputed mercury-sensitive genotype of *Gpi*-2 in mosquitofish, detected from bioassays and field studies, was more resistant to enzyme inhibition from mercury than other genotypes (Kramer et al. 1992a; Kramer and Newman 1994). The mercury concentration resulting in 50% inhibition of control activity (IC50) was greater in the reputed sensitive genotype than in a more resistant genotype, despite a greater uninhibited reaction velocity for the more resistant *Gpi*-2 genotype (Kramer and Newman 1994). Therefore, differential sensitivity to toxicity of heavy metals by different allozyme genotypes may not be mediated through differential enzyme inhibition.

Although differential enzyme inhibition was not detected, differential metabolic responses by different allozyme genotypes have been reported. DiMichelle et al. (1991) found that mummichogs with different allozymes of *Ldh*-B varied in their glycolytic metabolic dynamics. Additionally, Kramer et al. (1992a) found that a mercury-sensitive *Gpi*-2 allozyme genotype in mosquitofish exhibited a greater Krebs cycle metabolic response to mercury than other genotypes. In general, the metabolic response to mercury exposure by *Gpi*-2 was to decrease glycolysis but increase Krebs cycle activity (Kramer et al. 1992b). Therefore, different allozyme genotypes may

**TABLE 3.** *A list of species and chemical toxicants for which individuals with different allozyme genotypes exhibited differential survival*

| Chemical | Taxon | Species | Enzyme | Reference |
|---|---|---|---|---|
| Mercury | Caddisfly | *Nectopsyche albida* | Pgm<br>Adh | Benton and Guttman<br>1992a,b |
| | Shrimp | *Palaemon elegans* | Pgm<br>Pgm | Nevo et al. 1981<br>Ben-Shlomo and Nevo<br>1988 |
| | Gastropods | *Cerithium scabridum* | Aat<br>Aat<br>Acph<br>Glydh<br>Pgm | Lavie and Nevo 1986a<br>Lavie and Nevo 1988 |
| | Mosquitofish | *Gambusia affinis* | Idh<br>Mdh<br>Gpi | Diamond et al. 1989<br>Mulvey et al. 1995 |
| | | *Gambusia holbrooki* | Gpi | Heagler et al. 1993 |
| Copper | Gastropods | *Monodonta species* | Gpi | Lavie and Nevo 1982 |
| | Stoneroller | *Campostoma anomalum* | Gpi<br>Mdh | Gillespie and Guttman<br>1989 |
| | Fathead minnow | *Pimephales promelas* | Gpi<br>Idhp<br>Mdh<br>Pgm | Schlueter et al. 1997 |
| | Mosquitofish | *Gambusia affinis* | Gpi<br>Idh | Chagnon and Guttman<br>1989b |
| Cadmium | Gastropods | *Cerithium scabridum* | Gpi<br>G6pdh<br>Aat<br>Acph<br>Glydh<br>Pgm | Lavie and Nevo 1986a,b<br>Lavie and Nevo 1988 |
| | Shrimp | *Palaemon elegans* | Pgm | Ben-Shlomo and Nevo<br>1988 |
| | Mosquitofish | *Gambusia affinis* | Gpi | Chagnon and Guttman<br>1989b |
| Arsenic | Mosquitofish | *Gambusia affinis* | Gpi<br>Fh | Newman et al. 1989 |
| Zinc | Gastropod | *Monodonta species* | Gpi | Lavie and Nevo 1982 |
| Chlorpyrifos | Mosquitofish | *Gambusia affinis* | Gpi | Hughes et al. 1991 |
| Organo-<br>phosphate | Juvenile Bream | *Abramis brama* | Po | Chuiko and Slynko 1995 |
| Detergents,<br>Crude oil | Gastropods | *Monodonta species* | Gpi | Lavie et al. 1984 |
| Increased<br>acidity &<br>aluminum | Central<br>Mudminnow | *Umbra limi* | Mdh | Kopp et al. 1992 |

**TABLE 4.** *List of studies that have associated allozyme variation with biochemical activities of enzymes*

| | | |
|---|---|---|
| *Fundulus heteroclitus* | Variation in metabolic rate was strongly associated with allozyme genotypes of *Ldh*-B | DiMichelle et al. 1991 |
| Fish species | Allozymes of *Pgm* in 17 species of fish varied in their in vitro threshold to inhibition by copper | Chagnon and Guttman 1989a |
| *Gambusia holbrooki* | Isoalleles of *Pgm* showed differential inhibition to copper | Chagnon and Guttman 1989a |
| | No evidence of differential inhibition of allozymes of *Gpi* and *Mdh* after exposure to mercury | Kramer and Newman 1994 |
| | Fish with genotype *Gpi*-2 38/38 had an overall increase in glycolytic activity in response to mercury exposure and were more sensitive to mercury toxicity than those with other *Gpi*-2 genotypes | Kramer et al. 1992a |
| Variety of species from bacteria to humans | Relative electrophoretic mobility of *Gpi* allozymes may have adaptive significance. The charge of an allelic form of *Gpi* may affect its function in metabolic pathways. Evidence suggests that *Gpi* may be a target of metal toxicity. Differences in allelic forms of *Gpi* may vary in their ability to resist inhibition by metals | Riddoch 1993 |

vary in their general response to the stress of metal exposure. The relationship between differential metabolic response of allozymes and differential sensitivity to toxicants is unknown and needs further study.

Although the mechanisms are unknown, toxicants may directly interact with specific allozyme genotypes to affect their activity. Metal inhibition of *Pgm* activity is well documented (Milstein 1961a,b), and electrophoretic mobility of *Gpi* has been proposed as an indicator of differences in adaptive potential in many species (Riddoch 1993; Watt 1977). When combined with known differential metabolic responses, it is possible that allozymes may be direct sites of action for toxicity to certain toxicants (e.g., metals) and variation may contribute to differential sensitivity to toxicity.

Although evidence from laboratory studies suggests that allozymes may be sites of action for chemical toxicants, we cannot exclude the possibility that allozymes are linked to other genetically-based processes that confer tolerance to toxicity. Because differences in resistance to toxicity can occur from variation at sites of action (e.g., receptors, enzymes) and modifiers of toxicity (metallothioneins, MFOs), genotypic sensitivity may be polygenic, and allozymes may be merely markers of a more multigenic resistance genotype. The consistency of enzyme loci that show differential survivorship among several species and the unlikely scenario that electrophoretically-detected enzymes are tightly linked to other tolerance-producing processes in so many different species argue against the linkage hypothesis. However, even if allozymes are only markers of a genetically-based tolerance to toxic effects of chemicals, they are still valuable for studies on the effects of pollutants on genetic variation in the field and in the laboratory.

## SIGNIFICANCE OF CHANGES IN ALLOZYME FREQUENCIES

Changes in allozyme frequencies may be associated with significant negative impacts to populations. It could be argued that a significant negative impact could only come from either a significant reduction in heterozygosity and polymorphism or from a reduction in fitness of a population. A loss of genetic variation and/or reduced fitness in a population may leave a species more vulnerable to extinction. Thus, studies that link chemical-induced changes in allozyme frequencies with a loss of heterozygosity, polymorphism, or reduced fitness would provide evidence that changes in allozyme frequencies are significant negative impacts.

### Allozyme Frequencies and Genetic Variation

Although the evidence is somewhat equivocal, a few studies have associated exposure to environmental chemical toxicants with changes in genetic variation (Table 5). Two studies reported that contaminant-exposed populations had greater heterozygosity than reference populations, five studies reported significantly less heterozygosity, and one case reported 50% less polymorphism of allozyme loci in populations exposed

**TABLE 5.** *List of studies that associated exposure to contaminants with polymorphism and heterozygosity in natural populations[1]*

| Species | Effects on polymorphism | Effects on heterozygosity | Number of alleles lost | Number of loci | Reference |
|---|---|---|---|---|---|
| Gambusia affinis | 50% less polymorphic[2] | Significantly less heterozygosity | 3 | 5 | Keklak et al. 1994 |
| Gambusia holbrooki | No data | Significantly less heterozygosity | 4 | 5 | Benton et al. 1994 |
| Pimephales notatus | No effect[3] | Significantly less heterozygosity[4] | — | 7 | Fore et al. 1995a |
| Umbra limi | No data | Significantly less heterozygosity | — | 8 | Kopp et al. 1992 |
| Luxilus cornutus | No data | Significantly less heterozygosity | — | 14 | Kopp et al. 1994 |
| Semotilus atromaculatus | No data | Significantly less heterozygosity | — | 14 | Kopp et al. 1994 |
| Campostoma anomalum | No effect[3] | Significantly greater heterozygosity[4] | — | 8 | Fore et al. 1995b |
| Gila seminuda | No data | Excess heterozygosity | — | 3 | DeMarais et al. 1993 |

[1] The list only includes studies in which at least three loci were variable. Dashed lines indicate that no data were collected.

[2] Percent polymorphic loci.

[3] Mean number of alleles per locus.

[4] Mean proportion of heterozygosity within individuals.

to environmental contaminants. Because the minimum amount of heterozygosity and polymorphism necessary for a population to survive is unknown, it may be best to consider any significant reduction in these parameters to be a negative impact. Less heterozygosity and polymorphism may indicate a loss of ability to adapt to further environmental changes.

Although standards for minimum genetic variation may not yet exist, many population genetic researchers have shown a positive relationship between heterozygosity and the ability of a population to adapt to changing environmental stressors (Mitton and Grant 1984; Allendorf and Leary 1986; Hedrick et al. 1976; Powers et al. 1991). Additionally, laboratory studies that exposed animals to chemical toxicants have shown that heterozygosity may be positively correlated with tolerance to the toxicity of environmental pollutants (Table 6). Although three studies showed neutral or opposite effects, four laboratory studies provided evidence that the tolerance of several species to the toxic effects of different chemical toxicants was positively correlated with heterozygosity of allozyme loci.

The evidence from some studies that environmental toxicants reduce heterozygosity in natural populations and that heterozygous individuals in the laboratory are more resistant to the effects of toxicity than homozygous individuals seems to conflict. However, selection for resistance in nature is long-term, and it would be expected that individuals with homozygous genotypes for resistant alleles would be more fit than heterozygotes. Therefore, when a population is exposed to chemical toxicants

**TABLE 6.** *List of laboratory studies that have associated differential survivorship of individuals with heterozygosity of allozyme genotypes*

| Species | Chemical toxicant | Effects on heterozygosity | Number of loci | Reference |
|---------|-------------------|---------------------------|----------------|-----------|
| Gambusia affinis | Mercury | No relationship with heterozygosity | 9 | Mulvey et al. 1995 |
| Gambusia affinis | Mercury | Increased tolerance positively correlated with heterozygosity | 6 | Diamond et al. 1989 |
| Gambusia affinis | Arsenic | Increased tolerance in males negatively correlated with heterozygosity | 8 | Newman et al. 1989 |
| Umbra limi | Acidity Aluminum | Tolerant individuals had significantly greater heterozygosity | 5 | Kopp et al. 1992 |
| Nectopsyche albida | Mercury | Increased tolerance positively correlated with heterozygosity | 4 | Benton and Guttman 1992b |
| Pimephales promelas | Copper | No relationship with heterozygosity | 7 | Schlueter et al. 1995, 1997 |
| Monodonta Littoria Cerithium | Heavy metals Detergents Crude oil | Species with greater heterozygosity were more tolerant | 17 26 25 | Nevo et al. 1986 |

over time, we would expect a tendency towards homozygosity at all loci that contribute to resistance and a trend of reduced heterozygosity. In the laboratory, however, short-term exposure of one generation does not mimic selection in nature. Heterozygotes of a single generation may simply have the greatest probability for having at least one resistant allele for all loci that contribute to tolerance and are therefore more resistant than homozygotes. Multigenerational studies in the laboratory are needed to show that exposure of populations to toxic chemicals over time will select for individuals that are homozygous for resistant alleles.

The effect of chemical toxicants on genetic variation in natural populations is an important relationship that needs further study. Although difficult to discern from other factors that affect allelic variation, field studies are needed that can accurately determine the effect of environmental contaminants on genetic variation in natural populations.

## Allozymes and Fitness in Natural Populations

It is difficult to closely link changes in allozyme frequencies with changes in fitness of a population. In fact, we could find only three studies that made associations between exposure to chemical pollutants, allozyme variation, and fitness parameters (Table 7). Interestingly, three studies reported that allozyme genotypes that were relatively resistant to chemical toxicity of metals were also associated with smaller relative size (Benton et al. 1994; Schlueter et al. 1995, 1997). Another study showed that one allozyme genotype of *Gpi* had less reproductive potential than other genotypes after exposure to mercury (Mulvey et al. 1995). Although no studies specifically associate environmental chemical pollutants with variation in fitness of allozyme genotypes, several others have shown that certain fitness characteristics vary among allozyme genotypes.

In studies that range in species from insects to mammals, a variety of fitness charac-teristics associated with allozyme genotype and/or heterozygosity included metabolic rates, developmental rates, growth rates, time to maturity, swimming speed, pheno-typic variation, and complex behavior (Table 7). Although there is a lack of direct evidence for chemical toxicity and fitness of allozyme variants, important fitness char-acteristics have been associated with allozyme variation in many species. Therefore, there is a possibility that environmental chemical pollutants that randomly or selec-tively eliminate certain allozyme genotypes may also significantly reduce the fitness of the population by eliminating alleles that could contribute to survival and enhanced reproduction under other environmental conditions.

The relationship between exposure to environmental chemical pollutants, changes in allozyme frequencies, and changes in fitness characteristics needs much more study. If individuals with resistant genotypes of allozymes have less potential for growth and reproduction than those with sensitive genotypes, then we should be able to link changes in allozyme genotype frequencies with population declines in nature.

The importance of genetic variation in a population is well understood, even though the minimum amount of variation necessary for a population to be "healthy" is not.

**TABLE 7.** *List of studies that have associated allozyme variation with differences in fitness characteristics*

| Species | Evidence for associations between allozyme variation and fitness characteristics | Reference |
|---|---|---|
| *Callosohruchus maculatus* | Differences in allozyme genotypes and allozyme variation were associated with differences in behaviors that included bean-size discrimination, egg laying distribution, and larval competitiveness | Berg and Mitchell 1993 |
| *Argopecten irradians* | Variation in *Gpi* genotypes was associated with variation in growth of somatic tissue and reproductive tissue | Krause and Bricelj 1995 |
| *Mercenaria mercenaria* | No significant correlation between allozyme heterozygosity and growth rate | Slattery et al. 1993 |
| | Positive correlation between heterozygosity and age in one population. | |
| *Mulinia lateralis* | Positive relationship between allozyme heterozygosity and growth | Garton et al. 1984 |
| | A negative relationship between routine metabolic rates and heterozygosity explains most of the relationship between variation in growth among individuals and allozyme variation | |
| *Thais haemastoma* | Positive correlation between heterozygosity and scope for growth under salinity stress. Variation in growth was explained mostly by variation in feeding rates | Garton 1984 |
| *Heliosoma trivolvis* | Snails with *Gpi*-BC genotypes were associated with greater size plasticity than other genotypes | Benton et al. 1994 |
| | The *Gpi*-C allele was higher in frequency at contaminated sites and more frequent in smaller individuals. Thus, there may be a trade-off between tolerance to metal toxicity and growth for the *C* allele | |
| *Fundulus heteroclitus* | Heterozygous individuals had less phenotypic variation than homozygotes | Mitton 1978 |
| | Metabolic rate of embryos homozygous for *Ldh*-$B^a$ was higher than that for embryos homozygous for *Ldh*-$B^b$ during 24 hours after fertilization | Paynter et al. 1991 |
| | Developmental rates were significantly correlated with allozyme genotype variation at *Gpi* and *Mdh* | DiMichelle et al. 1986 |
| | Fish with genotypes of *Ldh*-$B^bB^b$ had a 20% faster swimming speed than those with *Ldh*-$B^aB^a$ at $10°C$. No differences were found at $25°C$ | DiMichelle and Powers 1982 |
| | Variation at four allozyme loci had an influence on developmental rate and thermal tolerance | DiMichelle and Powers |
| | Heterozygosity was not associated with increased rate of development nor thermal tolerance | 1991 |
| *Gambusia affinis* | Genotypes of *Gpi*-$2^{100/100}$ were only 43% gravid, whereas other genotypes were 70% gravid | Mulvey et al. 1995 |
| | Genotypes of *Gpi*-$2^{100/100}$ had significantly less embryos per female than other genotypes | |
| | *Gpi*-$2^{100/100}$ was also the most common allozyme genotype and sensitive to mercury toxicity | |

*(Continued)*

**TABLE 7.** *Continued*

| Species | Evidence for associations between allozyme variation and fitness characteristics | Reference |
|---|---|---|
| | Heterozygosity was positively correlated with growth especially at high temperatures | Mulvey et al. 1994 |
| | Significant differences in time to maturity were associated with allozyme genotype | |
| | Heterozygosity was negatively related to fluctuating asymmetry | |
| *Pimephales promelas* | Fish with different *Idhp* and *Mdh* genotypes showed differential survivorship to copper toxicity and growth rate | Schlueter et al. 1995, 1997 |
| | Smaller fish had more copper-resistant allozyme genotypes than larger fish | |
| *Salvelinus fontinalis* | Heterozygosity was not correlated with life history traits | Hutchings and Ferguson 1992 |
| *Cottus species* | Significantly negative correlation between individual heterozygosity and relative morphological variability | Strauss 1991 |
| *Salmo salar* | Populations with higher frequencies of *Pgm*-1r* had higher frequencies of precociously mature males | Pollard et al. 1994 |
| *Zonotrichia capensis* | Significant negative correlation between *individual* heterozygosity and morphological variation | Yezerinac et al. 1992 |
| | A weak positive correlation between *population* heterozygosity and morphological variation | |
| *Peromyscus polionotus* | Mean body weight was significantly positively correlated with individual heterozygosity in mainland populations, but not in beach populations | Garten 1976 |
| | Levels of aggressiveness were positively correlated with heterozygosity in mainland populations, but not in beach populations | |

Although the effects of allozyme variation on fitness in a natural population will be difficult to study, it is important to know what this relationship is under various contaminant conditions. Until we better understand this relationship, the significance of changes in allozyme frequencies will remain unknown.

## CONCLUSIONS

Our objective was to present evidence that environmental chemical contaminants are associated with changes in allozyme frequencies in natural populations. Additionally, we hoped to show evidence that these changes can produce negative impacts on populations. Studies suggest that variations in allozyme frequencies of natural populations are associated with exposure to chemical toxicants and that individuals with different allozyme genotypes have differential tolerances to the toxic effects of chemicals. However, the evidence that changes in allozyme frequencies produce significant negative impacts is less grounded. Although there is evidence that changes in allozyme frequencies can result in a loss of heterozygosity, polymorphism, and fitness, the support for this hypothesis is only indirect.

In order to rigorously test the hypothesis that contaminant-induced alterations in allozyme variation produce negative impacts in populations, more holistic studies are needed (Guttman 1994). Several field sites with different environmental exposure conditions should be selected and several species should be identified as model taxa. Allozyme analysis should be used on many species to detect differences in polymorphism and heterozygosity that can be correlated with environmental pollutants. Once the possible scenario of the relationship between chemical toxicants and allozyme changes are identified, laboratory studies should be used to confirm that changes in allozyme frequencies are due to chemical exposure. Finally, growth and reproductive output should be measured and compared among individuals with sensitive and tolerant allozyme genotypes. Although changes in allozyme frequencies could be caused by exposure to toxicants, the significance of these changes can be objectively assessed only after we know their impact on the susceptibility of a population to extinction.

Because there is a need to better understand the relationship between exposure to chemical pollutants, genetic variation, and negative impacts to ecosystems, researchers have proposed a new science of "Genetic Ecotoxicology" (Anderson et al. 1994). This discipline could study the full range of effects that anthropogenic environmental stressors have on genes, gene products and gene frequencies. However, because this new science is in its infancy (Evenden and Depledge 1997) and much is yet to be done, perhaps it would be wise to at least use allozyme analysis to monitor changes in genetic variation of natural populations.

The stability of ecosystems may very well depend upon the ability of populations to adapt. Furthermore, the adaptive potential of a population may very well depend upon the genetic variation it possesses. Therefore, as proposed by Fox (1995), monitoring genetic variation in natural populations may help us protect "the tinkerer" by protecting the "parts" of the ecological machinery whose functions we have yet to understand.

## REFERENCES

Allendorf, F. W., and R. F. Leary. 1986. Heterozygosity and fitness in natural populations of animals. In *Conservation biology: The science of scarcity and diversity*, ed. M. E. Soule, 57–76. Sunderland, MA: Sinauer Associates.

Anderson, S., W. Sadinski, L. Shugart, P. Brussard, M. Depledge, T. Ford, J. Hose, J. Stegeman, W. Suk, I. Wirgin, and G. Wogan. 1994. Genetic and molecular ecotoxicology: A research framework. *Environmental Health Perspectives* 102(Suppl):3–8.

Avise, J. C. 1994. *Molecular markers, natural history and evolution*. New York: Chapman & Hall.

Baird, D. J., I. Barber, M. Bradley, A. M. Soares, and P. Calow. 1991. A comparative study of genotype sensitivity to acute toxic stress using clones of *Daphnia magna* Straus. *Ecotoxicology and Environmental Safety* 21:257–265.

Baird, D. J., I. Barber, and P. Calow. 1990. Clonal variation in general responses of *Daphnia magna* Straus to toxic stress. I. Chronic life-history effects. *Functional Ecology* 4:399–407.

Ben-Shlomo, R., and E. Nevo. 1988. Isozyme polymorphism as monitoring of marine environments: the interactive effect of cadmium and mercury pollution on the shrimp, *Palaemon elegans*. *Marine Pollution Bulletin* 19:314–317.

Benton, M. J., and S. I. Guttman. 1992a. Allozyme genotype and differential resistance to mercury pollution in the caddisfly, *Nectopsyche albida*. I. Single-locus genotypes. *Canadian Journal of Fisheries and Aquatic Sciences* 49:142–146.

Benton, M. J., and S. I. Guttman. 1992b. Allozyme genotype and differential resistance to mercury pollution in the caddisfly, *Nectopsyche albida*. II. Multilocus genotypes. *Canadian Journal of Fisheries and Aquatic Sciences* 49:147–149.

Benton, M. J., S. A. Diamond, and S. I. Guttman. 1994. A genetic and morphometric comparison of *Helisoma trivolvis* and *Gambusia holbrooki* from clean and contaminated habitats. *Ecotoxicology and Environmental Safety* 29:20–37.

Berg, D. J., and R. Mitchell. 1993. Associations of allozyme variation and behavior in the cowpea weevil (*Callosobruchus maculatus*). *Entomologia Experimentalis et Applicata* 69:215–220.

Chagnon, N. L., and S. I. Guttman. 1989a. Biochemical analysis of allozyme copper and cadmium tolerance in fish using starch gel electrophoresis. *Environmental Toxicology and Chemistry* 8:1141–1147.

Chagnon, N. L., and S. I. Guttman. 1989b. Differential survivorship of allozyme genotypes in mosquitofish populations exposed to copper or cadmium. *Environmental Toxicology and Chemistry* 8:319–326.

Chuiko, G. M., and Y. V. Slynko. 1995. Relation of allozyme genotype to survivorship of juvenile bream, *Abramis brama* L., acutely exposed to DDVP, an organophosphorous pesticide. *Bulletin of Environmental Contamination and Toxicology* 55:738–745.

DeMarais, B. D., T. E. Dowling, and W. L. Minckley. 1993. Post-perturbation genetic changes in populations of endangered Virgin River chubs. *Conservation Biology* 7:334–341.

Diamond, S. A., M. C. Newman, M. C. Mulvey, P. M. Dixon, and D. Martinson. 1989. Allozyme genotypes and time to death of mosquitofish, *Gambusia affinis* (Baird and Girard), during acute exposure to inorganic mercury. *Environmental Toxicology and Chemistry* 8:613–622.

DiMichelle, L., and D. A. Powers. 1982. Physiological basis for swimming endurance differences between LDH-B genotypes of *Fundulus heteroclitus*. *Science* 216:1014–1016.

DiMichelle, L., and D. A. Powers. 1991. Allozyme variation, developmental rate, and differential mortality in the Teleost *Fundulus heteroclitus*. *Physiological Zoology* 64:1426–1443.

DiMichelle, L., D. A. Powers, and J. A. DiMichelle. 1986. Developmental and physiological consequences of genetic variation of enzyme synthesizing loci in *Fundulus heteroclitus*. *American Zoologist* 26:201–208.

DiMichelle, L., K. T. Paynter, and D. A. Powers. 1991. Evidence of lactate dehydrogenase-B allozyme effects in the teleost, *Fundulus heteroclitus*. *Science* 253:898–900.

Eanes, W. F. 1987. Allozymes and fitness: Evolution of a problem. *Trends in Ecology and Evolution* 2:43–48.

Evenden, A. J., and M. H. Depledge. 1997. Genetic susceptibility in ecosystems: The challenge for ecotoxicology. *Environmental Health Perspectives* 105(Suppl 4):849–854.

Facemire, C. F., T. S. Gross, and L. J. Guillette, Jr. 1995. Reproductive impairment in the Florida panther: Nature or nurture? *Environmental Health Perspectives* 103(Suppl 4):79–86.

Fore, S. A., S. I. Guttman, A. J. Bailer, D. J. Altfater, and B. V. Counts. 1995a. Exploratory analysis of population genetic assessment as a water quality indicator. I. *Pimephales notatus*. *Ecotoxicology and Environmental Safety* 30:24–35.

Fore, S. A., S. I. Guttman, A. J. Bailer, D. J. Altfater, and B. V. Counts. 1995b. Exploratory analysis of population genetic assessment as a water quality indicator. II. *Campostoma anomalum*. *Ecotoxicology and Environmental Safety* 30:36–46.

Fox, G. 1995. Tinkering with the tinkerer: Pollution versus evolution. *Environmental Health Perspectives* 103(Suppl 4):93–100.

ffrench-Constant, R. H., J. C. Steichen, T. A. Rocheleau, K. Aronstein, and R. T. Roush. 1993. A single-amino acid substitution when a y-aminobutyric acid subtype A receptor locus is associated with cyclodiene insecticide resistance in *Drosophila* populations. *Proceedings of the National Academy of Science USA* 90:1957–1961.

Garten, C. T. 1976. Relationships between aggressive behavior and genic heterozygosity in the oldfield mouse, *Peromyscus polionotus*. *Evolution* 30:59–72.

Garton, D. W. 1984. Relationship between multiple locus heterozygosity and physiological energetics of growth in the estuarine gastropod *Thais haemastoma*. *Physiological Zoology* 57:530–543.

Garton, D. W., R. K. Koehn, and T. M. Scott. 1984. Multiple-locus heterozygosity and the physiological energetics of growth in the coot clam, *Mulina lateralis*, from a natural population. *Genetics* 108:445–455.

Gillespie, R. B. 1996. Allozyme frequency variation as an indicator of contaminant-induced impacts in aquatic populations. In *Techniques in aquatic toxicology*, ed. G. K. Ostrander, 247–275. Boca Raton, FL: Lewis Publishers.

Gillespie, R. B., and S. I. Guttman. 1989. Effects of contaminants on the frequencies of allozymes in populations of the central stoneroller. *Environmental Toxicology and Chemistry* 8:309–317.

Gillespie, R. B., and S. I. Guttman. 1993a. Allozyme frequency analysis of aquatic populations as an indicator of contaminant-induced impacts. In *Environmental toxicology and risk assessment*: 2nd volume, ASTM STP 1173, eds. J. Gorsuch, F. J. Dwyer, C. G. Ingersoll, and T. W. LaPointe, 134–145. Philadelphia: American Society for Testing and Materials.

Gillespie, R. B., and S. I. Guttman. 1993b. Correlations between water quality and frequencies of allozyme genotypes in spotfin shiner (*Notropis spilopterus*) populations. *Environmental Pollution* 81:147–150.

Graves, J. E., and G. N. Someio. 1982. Electrophoretic and functional enzymic evolution in four species of Eastern Pacific barracudas from different thermal environments. *Evolution* 36:97–106.

Guttman, S. I. 1994. Population genetic structure and ecotoxicology. *Environmental Health Perspectives* 102(Suppl):97–100.

Heagler, M. G., M. C. Newman, M. Mulvey, and P. M. Dixon. 1993. Allozyme genotype in mosquitofish, *Gambusia holbrooki*, during mercury exposure: Temporal stability, concentration effects and field verification. *Environmental Toxicology and Chemistry* 12:385–395.

Hedrick, P. W., M. E. Ginevan, and E. P. Ewing. 1976. Genetic polymorphism in heterogeneous environments. *Annual Review of Ecology and Systematics* 7:1–32.

Holloway, G. J., R. M. Sibly, and S. R. Povey. 1990. Evolution in toxin-stressed environments. *Functional Ecology* 4:289–294.

Hughes, J. M., D. A. Harrison, and J. M. Arthur. 1991. Genetic variation at the PGI locus in the mosquito fish *Gambusia affinis* (Poeciliidae) and a possible effect on susceptibility to an insecticide. *Biological Journal of the Linnean Society* 44:153–167.

Hutchings, J. A., and M. M. Ferguson. 1992. The independence of enzyme heterozygosity and life-history traits in natural populations of *Salvelinus fontinalis* (brook trout). *Heredity* 69:496–502.

Kasai, F., and T. Hanazato. 1995. Genetic changes in phytoplankton communities exposed to the herbicide Simetryn in outdoor experimental ponds. *Archives of Environmental Contamination and Toxicology* 28:154–160.

Keklak, M. M., M. C. Newman, and M. Mulvey. 1994. Enhanced uranium tolerance of an exposed population of the eastern mosquitofish (*Gambusia holbrooki* Girard 1859). *Archives of Environmental Contamination and Toxicology* 27:20–24.

Kimura, M. 1993. Retrospective of the last quarter century of the neutral theory. *Japanese Journal of Genetics* 68:521–528.

Klerks, P. L., and J. S. Levinton. 1989. Effects of heavy metals in a polluted aquatic ecosystem. In *Ecotoxicology: Problems and approaches*, eds. S. A. Levin, M. A. Harwell, J. R. Kelly, and K. D. Kimball, 41–67. New York: Springer-Verlag.

Kopp, R. L., S. I. Guttman, and T. E. Wissing. 1992. Genetic indicators of environmental stress in central mudminnow (*Umbra limi*) populations exposed to acid deposition in the Adirondack Mountains. *Environmental Toxicology and Chemistry* 11:665–676.

Kopp, R., T. E. Wissing, and S. I. Guttman. 1994. Genetic indicators of environmental tolerance among fish populations exposed to acid deposition. *Biochemical Systematics and Ecology* 22:459–475.

Kramer, V. C., and M. C. Newman. 1994. Inhibition of glucosephosphate isomerase allozymes of the mosquitofish, *Gambusia holbrooki*, by mercury. *Environmental Toxicology and Chemistry* 13:9–14.

Kramer, V. J., M. C. Newman, M. Mulvey, and G. R. Ultsch. 1992a. Glycolysis and Krebs cycle metabolites in mosquitofish, *Gambusia holbrooki*, Girard 1859, exposed to mercuric chloride: Allozyme genotype effects. *Environmental Toxicology and Chemistry* 11:357–364.

Kramer, V. J., M. C. Newman, and G. R. Ultsch. 1992b. Changes in concentrations of glycolysis and Krebs cycle metabolites in mosquitofish, *Gambusia holbrooki*, induced by mercuric chloride and starvation. *Environmental Biology of Fishes* 34:315–320.

Krause, M. K., and V. M. Bricelj. 1995. Gpi genotypic effect on quantitative traits in the northern bay scallop, *Argopecten irradians irradians*. *Marine Biology* 123:511–522.

Lande, R. 1988. Genetics and demography in biological conservation. *Science* 241:1455–1460.

Lavie, B., and E. Nevo. 1982. Heavy metal selection of phosphoglucose isomerase allozymes in marine gastropods. *Marine Biology* 71:17–22.

Lavie, B., and E. Nevo. 1986a. The interactive effects of cadmium and mercury pollution on allozyme polymorphisms in the marine gastropod *Cerithium scabridum. Marine Pollution Bulletin* 17: 21–23.

Lavie, B., and E. Nevo. 1986b. Genetic selection of homozygote allozyme genotypes in marine gastropods exposed to cadmium pollution. *Science of the Total Environment* 57:91–98.

Lavie, B., and E. Nevo. 1988. Multilocus genetic resistance and susceptibility to mercury and cadmium pollution in the marine gastropod, *Cerithium scabridum. Aquatic Toxicology* 13:291–296.

Lavie, B., E. Nevo, and Y. Zoller. 1984. Differential viability of phosphoglucose isomerase allozyme genotypes of marine snails in nonionic detergent and crude oil mixtures. *Environmental Research* 35:270–276.

Lewin, R. 1987. The surprising genetics of bottlenecked flies. *Science* 235:1325–1327.

Milstein, C. 1961a. On the mechanism of activation of phosphoglucomutase by metal ions. *Biochemical Journal* 79:574–584.

Milstein, C. 1961b. Inhibition of phosphoglucomutase by trace metals. *Biochemical Journal* 79:591–596.

Mitton, J. B. 1978. Relationship between heterozygosity for enzyme loci and variation of morphological characters in natural populations. *Nature* 273:661–662.

Mitton, J. B., and M. C. Grant. 1984. Associations among protein heterozygosity, growth rate and developmental homeostasis. *Annual Review of Ecology and Systematics* 15:479–499.

Mitton, J. B., and R. K. Koehn. 1975. Genetic organization and adaptive response of allozymes to ecological variables in *Fundulus heteroclitus. Genetics* 79:97–111.

Mizrahi, L., and Y. Achituv. 1989. Effect of heavy metal ions on enzyme activity in the Mediterranean mussel, *Donax trunculus. Bulletin of Environmental Contamination and Toxicology* 42:854–859.

Morton, R. A. 1993. Evolution of *Drosophila* insecticide resistance. *Genome* 36:1–7.

Mouches, C., N. Pasteur, J. B. Berge, O. Hyrien, M. Raymond, B. R. De Saint Vincent, M. De Silvestri, and G. P. Georghious. 1986. Amplification of an esterase gene is responsible for insecticide resistance in a California *Culex* mosquito. *Science* 233:778–779.

Mulvey, M., and S. A. Diamond. 1991. Genetic factors and tolerance acquisition in populations exposed to metals and metalloids. In *Metal ecotoxicology: Concepts and applications*, ed. M. C. Newman, 301–321. Chelsea, MI: Lewis Publishers.

Mulvey, M., G. P. Keller, and G. Meffe. 1994. Single and multiple-locus genotypes and life-history responses of *Gambusia holbrooki* reared at two temperatures. *Evolution* 48:1810–1819.

Mulvey, M., M. C. Newman, A. Chazal, M. M. Keklak, M. G. Heagler, and L. S. Hales, Jr. 1995. Genetic and demographic responses of mosquitofish (*Gambusia holbrooki* Girard 1859) populations stressed by mercury. *Environmental Toxicology and Chemistry* 14:1411–1418.

Murdoch, M. H., and P. D. N. Hebert. 1994. Mitochondrial DNA diversity of brown bullhead from contaminated and relatively pristine sites in the Great Lakes. *Environmental Toxicology and Chemistry* 13:1281–1289.

Nevo, B., T. Perl, A. Beiles, and D. Wool. 1981. Mercury selection of allozyme genotypes in shrimps. *Experientia* 37:1152–1154.

Nevo, B., R. Ben-Shlomo, and B. Lavie. 1984. Mercury selection of allozymes in marine organisms: Predictions and verification in nature. *Proceedings of the National Academy of Science USA* 81:1258–1259.

Nevo, B., R. Noy, B. Lavie, A. Beiles, and S. Muchtar. 1986. Genetic diversity and resistance to marine pollution. *Biological Journal of the Linnean Society* 29:139–144.

Nevo, E., T. Shimony, and M. Libini. 1977. Thermal selection of allozyme polymorphisms in barnacles. *Nature* 267:699–701.

Newman, M. C., S. A. Diamond, M. Mulvey, and P. Dixon. 1989. Allozyme genotype and time to death of mosquitofish, *Gambusia affinis* (Baird and Girard), during acute toxicant exposure: a comparison of arsenate and inorganic mercury. *Aquatic Toxicology* 15:141–156.

O'Brien, S. J., M. E. Roelke, L. Marker, A. Newman, C. A. Winkler, D. Meltzer, L. Colly, J. F. Evermann, M. Bush, and D. E. Wildt. 1987. Genetic basis for species vulnerability in the cheetah. *Science* 227:1428–1434.

Paynter, K. T., L. DiMichelle, S. C. Hand, and D. A. Powers. 1991. Metabolic implications of *Ldh*-B genotype during early development in *Fundulus heteroclitus. Journal of Experimental Zoology* 257:24–33.

Pierce, B. A., and D. K. Wooten. 1992. Genetic variation in tolerance of amphibians to low pH. *Journal of Herpetology* 26:422–429.

Pollard, S. M., R. G. Danzmann, and R. R. Claytor. 1994. Association between the regulatory locus Pgm-1r* and life history types of juvenile Atlantic salmon (*Salmo salar*). *Canadian Journal of Fisheries and Aquatic Sciences* 51:1322–1329.

Powers, D. A., T. Lauerman, D. Crawford, and L. DiMichelle. 1991. Genetic mechanisms for adapting to a changing environment. *Annual Review of Genetics* 25:629–659.

Riddoch, B. J. 1993. The adaptive significance of electrophoretic mobility in phosphoglucose isomerase (PGI). *Biological Journal of the Linnean Society* 50:1–17.

Schlueter, M. A., S. I. Guttman, J. T. Oris, and A. J. Bailer. 1995. Survival of copper-exposed juvenile fathead minnows (*Pimephales promelas*) differs among allozyme genotypes. *Environmental Toxicology and Chemistry* 14:1727–1734.

Schlueter, M. A., S. I. Guttman, J. T. Oris, and A. J. Bailer. 1997. Differential survival of fathead minnows, *Pimephales promelas*, as affected by copper exposure, prior population stress, and allozyme genotypes. *Environmental Toxicology and Chemistry* 16:939–947.

Schat, H., E. Kuiper, W. M. Ten Brookum, and R. Voos. 1993. A general model for the genetic control of copper tolerance in *Silene vulgaris*: Evidence from crosses between plants from different tolerant populations. *Heredity* 70:142–147.

Slattery, J. P., R. A. Lutz, and R. C. Vrijenhoek. 1993. Repeatability of correlations between heterozygosity, growth and survival in a natural population of the hard clam *Mercenaria mercenaria*. *Journal of Experimental Marine Biology and Ecology* 165:209–224.

Smith, M. H., S. L. Scott, E. H. Liu, and J. C. Jones. 1983. Rapid evolution in a post-thermal environment. *Copeia* 1983:193–197.

Strauss, R. E. 1991. Correlations between heterozygosity and phenotypic variability in *Cottus* (Teleostei: Cottidae): character components. *Evolution* 45:1950–1956.

Strit, tholt, J. R., S. I. Guttman, and T. E. Wissing. 1988. Low levels of genetic variability of yellow perch (*Perca flavescens*) in Lake Erie and selected impoundments. In *The biogeography of the island region of Western Lake Erie*, ed. J. F. Downhower, 246–257. Columbus, OH: Ohio State University Press.

Stuhlbacher, A., M. C. Bradley, C. Naylor, and P. Calow. 1993. Variation in the development of cadmium resistance in *Daphnia magna* Straus: Effect of temperature, nutrition, age and genotype. *Environmental Pollution* 80:153–158.

Tranvik, L., M. Sjogren, and G. Bengtsson. 1994. Allozyme polymorphism and protein profile in *Orchesella bifasciata* (Collembola): Indicative of extended metal pollution? *Biochemical Systematics and Ecology* 22:13–23.

van Beneden, R. J., and D. A. Powers. 1989. Structural and functional differentiation of two clinally distributed glucosephosphate isomerase allelic isozymes from the teleost *Fundulus heteroclitus*. *Molecular Biology and Evolution* 6:155–170.

Watt, W. B. 1977. Adaptation at specific loci. I. Natural selection on phosphoglucose isomerase of *Colias* butterflies: Biochemical and population aspects. *Genetics* 87:177–194.

Watt, W. B. 1994. Allozymes in evolutionary genetics: Self-imposed burden or extraordinary tool? *Genetics* 136:11–16.

Yezerinac, S. M., S. C. Lougheed, and P. Handford. 1992. Morphological variability and enzyme heterozygosity: Individual and population level correlations. *Evolution* 46:1959–1964.

*Genetics and Ecotoxicology*
Edited by V. E. Forbes
Copyright © 1999 Taylor & Francis

# 5

# Adaptation to Metals in the Midge *Chironomus riparius*: A Case Study in the River Dommel

Jaap F. Postma and Dick Groenendijk

**Abstract.** Several invertebrate species, able to survive and reproduce in metal contaminated areas, have been observed to be less sensitive to metals than their unexposed conspecifics. This is often interpreted as an indication of genetic adaptation to increased metal concentrations, but physiological acclimation cannot be ruled out in many cases. An ideal demonstration of genetic adaptation to metals should prove heritability of characteristics and provide evidence for selection for metal tolerance. Furthermore, alterations in selective pressure due to gene flow from surrounding nonadapted populations may be of importance. The aim of the present investigation was therefore to analyse an evolving metal adaptation in the midge *Chironomus riparius* (Diptera; Meigen 1804), and to estimate the influx of larval chironomids from surrounding nonadapted midge populations.

Experiments using laboratory-reared offspring ($F_1$) demonstrated that midge populations naturally exposed to high metal concentrations were genetically adapted to cadmium. However, the increased cadmium-tolerance in *C. riparius* also caused some negative side effects, including increased mortality and larval development time under control conditions. These differences in life history between metal adapted and reference populations were caused partly by an altered regulation of essential metals such as zinc. However, results of these experiments also indicated that gene flow from unadapted populations might play an additional role, especially because in situ population dynamics of metal-exposed and unexposed populations (located upstream) were almost identical. Additional experiments further demonstrated that the actual level of cadmium tolerance strongly fluctuated under field conditions. It is therefore concluded that gene flow of nonadapted larvae is an important structuring factor in the process of metal adaptation, the influence of which varies through the seasons in relation to water discharge and population densities. The resulting dynamic interaction of seasonal population dynamics and metal tolerance contrasts most other examples of metal adaptation in invertebrates, in which the selective pressure is rather constant and dispersal is probably limited. This study adds to the very few case studies on aquatic invertebrates for which conclusive evidence of genetic adaptation to metals has been given.

**Keywords.** Cadmium; *Chironomus riparius*; gene flow; genetic adaptation; physiological acclimation; population dynamics.

## INTRODUCTION

Populations of several invertebrates living in metal contaminated areas are less sensitive to metals than previously unexposed populations (reviewed by Klerks and Weis 1987 and Posthuma and van Straalen 1993). These results are often interpreted as an indication of genetic adaptation to increased metal concentrations, although physiological acclimation cannot be ruled out in many cases. According to Posthuma and van Straalen (1993, and adapted from Brandon 1990), an ideal demonstration of genetic adaptation to metals should consist of the following components:

1. Evidence that selection for metal tolerance has occurred.
2. Demonstration that, based on life-history characteristics, some individuals are better adapted than others.
3. Evidence that the characters involved in metal tolerance are heritable.
4. Information on the possible influence of gene flow.
5. Phylogenetic information to enable distinction between the original and the derived state of the character.

The number of studies on genetic adaptation to metals in invertebrates that meet these criteria is limited. Among terrestrial invertebrates, metal adaptation has been conclusively demonstrated for five species only (Posthuma and van Straalen 1993): the fruitfly *Drosophila melanogaster*, the isopod *Porcellio scaber*, and three springtail species *Orchesella cincta*, *Isotoma notabilis*, and *Onychiurus armatus*. Among aquatic invertebrates, such studies are even more scarce, as only three cases are known: Brown (1976, the isopod *Asellus meridianus*), Nevo et al. (1984, the gastropod *Monodonta turbinata*), and Klerks and Levinton (1989, the oligochaete *Limnodrilus hoffmeisteri*).

However, even these studies did not meet all criteria mentioned. The life-history consequences of adaptation have, for example, not always been verified, and not all experiments conclusively demonstrated that selection for tolerance indeed had occurred. More important, however, is the general lack of information concerning the fourth and fifth criteria, in particular the influence of gene flow from surrounding nonadapted populations. This might be due to the problems usually encountered when trying to quantify the role of gene flow. However, in the case of river dwelling chironomids, gene flow is relatively easy to estimate because drifting larvae normally represent a major part of the total gene flow as compared to the dispersal of the short living and weak flying imagines (cf. Davies 1976).

The aim of the present investigation was therefore to assess the presence and evolution of metal adaptation in the midge *C. riparius* (Diptera; Meigen, 1804) and the possible role of gene flow. After a description of the research area, the results of published and current studies shall be considered in terms of the five criteria mentioned.

## THE RESEARCH AREA

Research on *C. riparius* was conducted in the River Dommel, a second to third order lowland river situated in northern Belgium (Fig. 1). The source of metal contamination in this river is a zinc factory of Union Minière on the banks of a small tributary of the Dommel, the Eindergatloop.

This factory started producing zinc and cadmium from zinc ores in 1888. Yearly production during the 1980s was on average 120,000 t zinc year$^{-1}$ and 600 t cadmium year$^{-1}$. Furthermore, a diffuse input of zinc in the catchment area is responsible for elevated background levels of zinc even at sites located upstream from the Eindergatloop (Table 1). Average yearly amounts of metals transported through the Dommel downstream from the Eindergatloop are 1–3 kg Cd day$^{-1}$ and 50–200 kg Zn day$^{-1}$, with a maximum water flow of 4.5 m$^3$ s$^{-1}$. The Dommel is further characterized by a sandy bottom, a width of 5–7 m, a depth of 0.4–1.5 m, a current velocity varying between 0.3 and 0.8 m s$^{-1}$, and neutral waters with a naturally high iron content.

For this study, several sampling locations in the Dommel were selected (Fig. 1). The upstream part of the river was sampled near the village Peer. Other major sampling locations were DEG, an unpolluted reference site just upstream from the Eindergatloop,

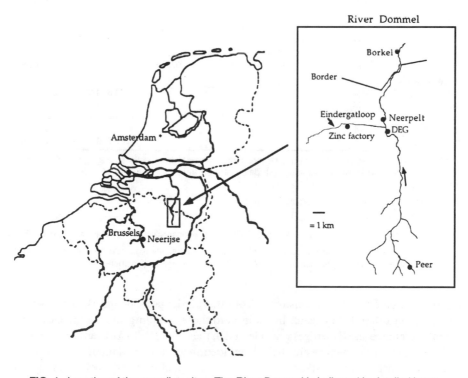

**FIG. 1.** Location of the sampling sites. The River Dommel is indicated in detailed inset.

**TABLE 1.** *Ranges of principal water characteristics and trace metal concentrations in water from the different field sites[a]*

| River station | | IJse | Dommel | | | |
| Parameter | Units | Neerijse | Peer | DEG | Neerpelt | Borkel |
|---|---|---|---|---|---|---|
| Width | m | 2–4 | 3–4 | 5–6 | 6–7 | 7–9 |
| Depth | m | 0.8–1.2 | 0.1–0.5 | 0.3–1.1 | 0.5–1.0 | 0.5–1.0 |
| Current velocity | m s$^{-1}$ | 0.22 | 0.58 | 0.27–0.77 | 0.83 | 0.65 |
| Water temperature | °C | 5–18 | 4–16 | 4–19 | 5–20 | 5–16 |
| Visibility | m | <0.2–0.6 | >0.5 | <0.2–1 | <0.2–1 | <0.2–1 |
| Oxygen saturation | mM | 0.15–0.36 | 0.22–0.36 | 0.1–0.33 | 0.03–0.33 | 0.16–0.29 |
| | % | 44–92 | 63–91 | 16–87 | 7–91 | 50–89 |
| pH | | 7.3–7.9 | 6.5–7.2 | 6.5 | 6.3–7.7 | 6.5–7.1 |
| Conductivity | $\mu$S cm$^{-1}$ | 555–805 | 280–480 | 393 | 220–960 | 357–1212 |
| Suspended matter | mg l$^{-1}$ | | | | | 10–26 |
| DOC[b] | mg l$^{-1}$ | 6.6 | 11.7 | 8.1–9.5 | 8.7 | 4.3–8.4 |
| $NO_3 + NO_2 - N$[c] | $\mu$M | 457–636 | 14–421 | | 78–921 | 214–1214 |
| $NH_4 - N$[c] | $\mu$M | 164–279 | 30–364 | | 64–900 | 59–400 |
| $PO_4 - P$[c] | $\mu$M | 6.5–38.7 | 16.1–35.5 | 9.3 | 9.7–171 | 3.2–51.6 |
| Cl[c] | mM | 2.1 | 1.2 | | 2.6 | 1.0–3.4 |
| $SO_4^c$ | mM | | | 0.7 | | 0.6–7.3 |
| Fe[c] | $\mu$M | 2.3–2.7 | 9.0–15.1 | 1.5–12.4 | 0.2–13.4 | 0.4–26.4 |
| Ca | mM | | | | | 0.9–1.1 |
| Mg | mM | | | | | 0.26–0.31 |
| K | mM | | | | | 0.19–0.56 |
| Na | mM | | | | | 0.78–3.74 |
| Trace metals[c] | | | | | | |
| Cd | nM | 0.3–3.4 | 4.5–12.5 | 0.9–12.1 | 24.6–2384* | 11.6–854* |
| Zn | $\mu$M | <0.30–0.9 | 3.5–4.6 | 0.9–4.0 | 0.9–68.8 | 2.1–17.7 |
| Cu | nM | 29.9–378 | 78.7–209 | 14.2–132 | 11.9–195 | 14.2–236 |
| Pb | nM | 1.9 | 2.9 | <0.9–4.8 | 8.2 | <0.9–3.9 |
| Trace metals[d] | | | | | | |
| Cd | nM | | | | 19.6–1838 | 20.5–2847 |
| Zn | $\mu$M | | | | 0.92–110 | 3.52–41.3 |
| Cu | nM | | | | | 63.0–598 |
| Pb | nM | | | | | 19.3–212 |
| Hg | nM | | | | <2.5–125 | 0.1–1.6 |

[a]Ranges are obtained by combining regular measurements by Water Quality Managing Board 'De Dommel' and the 'Vlaamse Milieumaatschappij' with data obtained from van Hattum et al. (1991) and our own observations.

[b]Concentrations in filtered water samples (0.2 $\mu$m).

[c]Concentrations in filtered water samples (0.45 $\mu$m).

[d]Total of particulate and dissolved trace metals. The number of observations varies between one and 30.

*Maximum Cd value is normally around 300 nM, but these extremely high values were recorded from January to July 1995. The reference site just upstream from the Eindergatloop is abbreviated as DEG.

Neerpelt, situated 500 m downstream from the Eindergatloop, and Borkel, located about 7 km further downstream. In addition to these sampling sites in the Dommel, another reference site (Neerijse) was selected in the River IJse at a distance of 100 km.

Although other midge species have been identified in the Dommel, *C. riparius* is by far the most dominant species at both reference and polluted sites. In the River Dommel, *C. riparius* is a multivoltine species, producing at least five generations per year (Groenendijk et al. 1998a), which is normal for this midge species in temperate climates. Emergence normally starts in March and intermittently continues until late October. Maximum larval densities are observed during late autumn and can be as high as 75,000 individuals m$^{-2}$. These high densities are recorded at reference and polluted sites.

## EVIDENCE THAT SELECTION HAS OCCURRED

Establishing that selection has occurred over the relevant time interval might be problematic. According to Brandon (1990), there are two basic reasons for this: 1) selection might have driven the selected trait to fixation and selection in action cannot be observed; 2) in cases where actual selection is observed, it should be extrapolated back through time and claimed that the present selective environment is similar to past selective environments. However, circumstantial evidence can be obtained from indications of present-day selection and the presence of other tolerant species. Posthuma and van Straalen (1993) concluded that combining data on the actual exposure in the field with dose-effect relationships for reference populations, including estimation of parameters such as EC$_{50}$ or NOEC (No Observed Effect Concentration), is crucial in obtaining circumstantial evidence for past selection. Comparing populations in a gradient around a point source can be informative, but attention should be paid to differences in habitat characteristics. In addition, information is needed for the extrapolation of laboratory-derived toxicity data to the actual exposure in the field. In the case of the Dommel and the midge *C. riparius*, circumstantial evidence based on the presence of other tolerant species is not available, although several other (midge) species inhabit polluted sites in the Dommel. Consequently, attention must be focused on either a comparison between the actual exposure in the field with dose-effect relationships obtained in the laboratory or on indications of present day selection.

### Actual Exposure in the Field

To aid the comparison of the actual exposure in the field with NOEC values obtained in the laboratory, chronic toxicity experiments with both cadmium and zinc were conducted, using a laboratory population of *C. riparius*, to assess effects on survival, larval growth, and reproduction (see Postma et al. 1994, 1995a for details). These experiments started with first instar larvae, less than 24 hours old, and ended after all surviving larvae had emerged. All treatments were tested in triplicate, and water was renewed once a week. At the onset of emergence the number of newly emerged males and females were sexed by examining the pupal exuviae. Furthermore, the number of deposited egg masses and the number of eggs per egg mass were also recorded daily. The hatchability of the egg masses was assessed after hatching in a petri dish at 20°C.

The results of these experiments, concerning the toxic effects of both cadmium and zinc on survival, larval growth, and reproduction are shown in Fig. 2 (Postma et al.

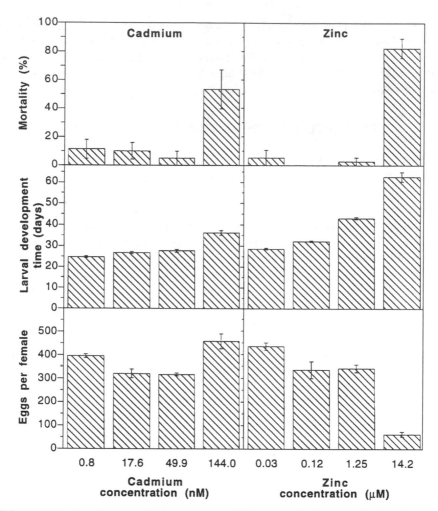

**FIG. 2.** Effects of cadmium and zinc exposure on *Chironomus riparius*: mortality, larval development ment time, and number of eggs per female. Mean values together with standard errors (±1 SE) are presented.

1994, 1995a, and based on actual metal concentrations in the water). Based on these toxicity data, as well as additional chronic toxicity data obtained from other studies (e.g., Pascoe et al. 1989; Timmermans et al. 1992), one generation NOEC values were estimated for several parameters: survival [50–140 nM Cd; 1.3–14 $\mu$M Zn], larval growth [20–50 nM Cd; <1.3$\mu$M Zn], and reproduction [>140 nM Cd; 0.12–1.25 $\mu$M Zn].

Lower NOEC values were, however, observed in multi-generation experiments because the toxic effects of cadmium significantly increased over the generations

(Postma and Davids 1995). In particular, the effects on larval growth increased over the generations, whereas the larval growth rate in the control was rather constant or only slightly increased. The NOEC value for larval growth will consequently be below the previously estimated 20 nM Cd. Whether this multi-generation effect also occurs with exposure to zinc is unknown. Miller and Hendricks (1996) maintained *C. riparius* cultures at increased zinc concentrations for 18 months but presented no data on the toxic effects of zinc for the separate generations produced during this period. Still, these experiments showed that tolerance to zinc increased over a long-term exposure to 0.15 $\mu$M. We, therefore, expect that field populations of *C. riparius* are negatively affected when cadmium or zinc concentrations in the water regularly exceed 20 nM Cd or 0.12 $\mu$M Zn.

Circumstantial evidence for selection in situ can be obtained by comparing these laboratory-derived toxicity data with measured concentrations in the field (Table 1). At polluted sites in the Dommel, NOEC values for both larval growth and even survival are regularly exceeded by a factor of 10–30 and 50–80 for cadmium and zinc, respectively. Even at upstream reference locations, estimated NOEC values for zinc were exceeded. Selection for tolerance is therefore likely to occur. However, toxicity in laboratory experiments is not necessarily identical to that in field situations, due to differences in, for example, metal speciation. Consequently, to determine whether or not the cadmium and zinc levels, which regularly exceed the NOEC values, negatively affect local chironomid populations, field studies or laboratory toxicity studies using surface water or sediment samples from contaminated locations must be conducted.

## Indications of Present Day Selection

Indications for present day selection were found in acute toxicity experiments, in which first instar larvae of *C. riparius*, obtained from a laboratory culture, were exposed to surface water obtained from several locations in the Dommel (Stuijfzand et al. 1996). These experiments demonstrated that when larvae were exposed to polluted surface water from Neerpelt, growth was about half that of larvae exposed to surface water from the upstream reference location DEG. Other indications were obtained in a detailed field study in which population dynamics at both reference and polluted sites were studied (Groenendijk et al. 1998a). This field study demonstrated that if sudden peaks in metal contamination coincide with peak densities of first or second instar larvae, high mortality can occur. Similar metal concentrations, on the other hand, resulted in little mortality when older instar larvae dominated the population.

In summary, it can be concluded that although evidence for selection in the past is missing, actual cadmium and zinc concentrations in the River Dommel are directly affecting local chironomid populations. Furthermore, seasonal variations in metal levels combined with seasonal occurrence of sensitive younger instar larvae are important factors for the resulting selective pressure and possibly for the resulting level of adaptation.

## DEMONSTRATION THAT, BASED ON LIFE-HISTORY CHARACTERISTICS, SOME INDIVIDUALS ARE BETTER ADAPTED THAN OTHERS

Differences between metal-exposed and unexposed populations can be detected by comparing life-history parameters such as survival, growth, or reproduction. Furthermore, insight into the physiological mechanisms causing an increased tolerance would benefit the interpretation of differences in life-history parameters. However, besides the direct selection by metals, several other factors can also influence life-history parameters. For example, selection by other habitat features might favor certain life-history characteristics unrelated to the increased metal tolerance (Posthuma and van Straalen 1993). The ultimate effect on life-history parameters will consequently depend on the strength of the separate factors and the amount of genetic variation available.

### Presence of Metal-Tolerant Chironomid Populations

Chronic toxicity experiments, using an experimental set-up similar to that described earlier, were conducted to compare life-history parameters between exposed (Neerpelt and Borkel) and unexposed populations (Peer and Neerijse, Fig. 3, based on Postma et al. 1995b). These experiments were started with unexposed laboratory-reared offspring ($F_1$) obtained from field-sampled larvae and demonstrated significant differences in the effects of cadmium on the four populations studied. Increased tolerance to cadmium in the two populations from metal contaminated sites was mainly expressed by a reduced effect on the larval growth rate while effects on larval mortality remained substantial. No interpopulation differences were found in the effects of cadmium on the reproduction of the midges (Fig. 3, Postma et al. 1995b).

The results of these experiments on cadmium-tolerance differed from experiments by Wentsel et al. (1978), who demonstrated that in addition to larval growth rates, the survival of chironomids obtained from exposed sites was less affected by metals compared with conspecifics obtained from a clean location. These contrasting results may be due to the use of field-sampled larvae (Wentsel et al. 1978) as opposed to larvae from the $F_1$-generation (present experiments). On the other hand, the genetic stability of the populations might also be a relevant variable, as will be discussed below.

### Physiological Mechanisms Causing an Increased Tolerance

Cadmium kinetics might differ between cadmium-adapted and nonadapted populations of *C. riparius*. Elimination experiments were started by using laboratory-reared offspring obtained from two exposed and one unexposed population (see Postma et al. 1996 for details). Larvae were cultured for a 16-day accumulation period in 176.2 nM Cd. In the fourth instar, 200 larvae were selected randomly from each population, and

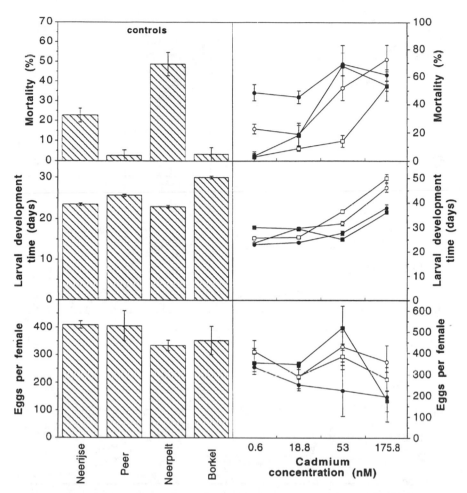

**FIG. 3.** Life-history parameters of two populations of *C. riparius* from exposed sites (closed symbols: Neerpelt = closed circles and Borkel = closed squares) and two from unexposed sites (open symbols: Neerijse = open circles and Peer = open squares). Presented are average values (together with standard errors (±1 SE)) on mortality, larval development time, and the number of eggs per female in both the controls (hatched bars) and the treatments with three different cadmium concentrations.

experiments were started by placing them in clean water. Some of the sampled larvae were dissected to analyse the guts (including Malpighian tubules) and the remainder of the larvae separately, whereas other larvae were frozen whole. A first-order one-compartment model was fitted to the experimental data to obtain estimates for the cadmium elimination rates. These experiments demonstrated that at all sampling times more than 90% of the cadmium was found in the guts, and elimination of cadmium by larvae closely resembled elimination from the guts. Furthermore, indications

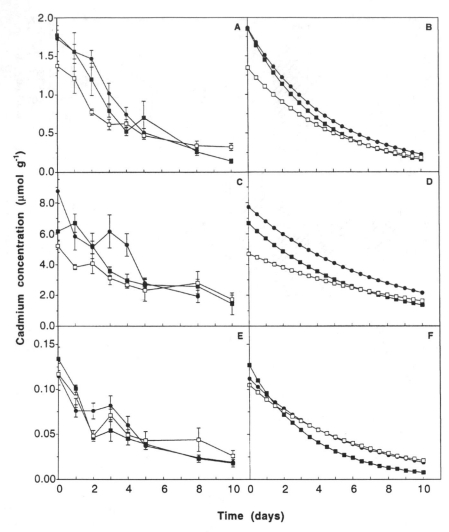

**FIG. 4.** Cadmium concentrations ($\mu$mol g$^{-1}$) in larvae from a non-adapted population (Peer = open squares) and two adapted populations (Neerpelt = closed circles and Borkel = closed squares) and the fitted one-compartment models during the elimination experiments: (*a*) larvae, data; (*b*) larvae, models; (*c*) guts, data; (*d*) guts, models; (*e*) larvae minus guts, data; ( *f* ) larvae minus guts, models. Mean values together with standard errors ($\pm$1 SE) are presented (reproduced with permission from Postma et al. 1996).

for an increased efficiency of cadmium excretion from the guts were found in both adapted populations, as compared to the reference population (Fig. 4, based on Postma et al. 1996). Such an increased excretion efficiency by the guts has also been found for metal-adapted collembola (Posthuma et al. 1992; van Straalen et al. 1987). In addition to the increased elimination rates, accumulation experiments demonstrated somewhat higher equilibrium values in cadmium-adapted larvae (10–20%; Postma et al. 1996). This increased storage capability could be due to the amount or the efficiency of metal-containing granula in the gut epithelium, because such concretions have been found in the midgut epithelium of several insects including chironomids (Lauverjat et al. 1989; Seidman et al. 1986b; Sohal and Lamb 1977). As these granula may be expelled into the gut lumen by exocytosis or degeneration of complete cells, they could provide a mechanism for increasing elimination efficiency. Indications of this were obtained by Seidman et al. (1986a), who found high amounts of cadmium in degenerating cells or debris from cells from the midgut epithelium of chironomids, which had been sloughed into the gut lumen. Other experiments demonstrated the induction of metallothionein-like proteins in metal-exposed chironomids (Seidman et al. 1986a; Yamamura et al. 1983). It has been suggested that these metal-loaded metallothioneins are picked up by granula in the midgut epithelium, where the metal can be retained after degradation of the protein (Dallinger 1993).

The increased tolerance to cadmium in the midge *C. riparius* may be due to a combination of several physiological mechanisms, which together cause an increased tolerance.

### Consequences of Increased Metal Tolerance for Life-History Patterns

As maintenance of a tolerance mechanism may be energetically expensive, "costs of tolerance" have frequently been suggested as an inevitable consequence of being tolerant (Holloway et al. 1990; cf. Shaw chapter 2). Indications for the presence of these costs can arise when tolerant individuals, cultured in a clean environment, have a lowered fitness compared to nontolerant individuals (Cook et al. 1972; Cox and Hutchinson 1981; Hickey and McNeilly 1975). However, as discussed by Posthuma and van Straalen (1993), adaptation of an exposed population to habitat characteristics other than metals or direct selection on a life-history parameter by the exposure may complicate these expectations. For example, life-history theory predicts that disturbances in a habitat resulting in reduced adult survival will select for earlier maturation and increased reproductive effort (Charlesworth 1980; Sibly and Calow 1989). For semelparous organisms like chironomids, the problem of finding the optimum life history is reduced to finding the optimal age of reproduction. Disturbances that reduce survival, such as metal pollution, may therefore tend to lower the age of reproduction (Gadgil and Bossert 1970).

The ultimate effect of these factors on life-history patterns of metal-tolerant chironomids will depend on the amount of genetic variation available, trade-offs among life-history characteristics and matings with nontolerant individuals (Falconer 1981; Mulvey and Diamond 1991). Experiments on cadmium-adapted chironomids cultured in the absence of cadmium (controls in Fig. 3, see Postma et al. 1995b for details) demonstrated that life-history patterns were unaffected by a direct selection for certain life-history characteristics because no indications were found for a lowered age of reproduction. This is in contrast to some terrestrial invertebrates, in which metal adapted populations have been characterised by an earlier maturation or an increased reproductive effort (Bajraktari et al. 1987; Donker et al. 1993; Posthuma et al. 1993b).

On the other hand, life-history patterns seemed to be affected by "costs of tolerance" because several experiments in which tolerant chironomids were cultured in a clean environment indicated a lowered fitness when compared to nontolerant individuals. The controls presented in Fig. 3, for example, demonstrated that both larval mortality (Neerpelt) and larval development time (Borkel) might be increased when compared with reference populations, where cadmium-tolerant midges were cultured in a clean environment. Also, such indications of "costs of tolerance" were found in experiments using a cadmium-adapted laboratory-reared population (Postma et al. 1995a). However, differences existed among cadmium-adapted populations. For example, control mortality among unexposed larvae from the adapted Neerpelt population and a cadmium-adapted laboratory-derived population was increased, whereas an increased mortality among unexposed larvae was absent in the adapted Borkel population. High mortality under clean conditions has also been found for metal-tolerant collembola (Posthuma et al. 1993b). A lack of metabolically available essential metals like zinc was suggested as one of the possible reasons, because an increased accumulation of zinc has been found in these organisms (Posthuma et al. 1992). An increased accumulation of zinc has also been found in chironomids (Postma et al. 1995a,b), but a causal relationship with the high control mortality seemed less likely because mortality remained high when cadmium-tolerant midges were supplied with additional zinc (Fig. 5; based on Postma et al. 1995a). The high control mortality most likely resulted from other (genetically determined) risks of tolerance development, which shall be discussed below.

These experiments, in which cadmium-tolerant midges were supplied with additional zinc, further demonstrated that the increase in larval development time, as observed in some tolerant chironomid populations when cultured under control conditions, was at least partly due to an increased need for essential metals, i.e., zinc (Postma et al. 1995a). Interpreting the reduced fitness of metal-adapted populations reared in a clean environment as "costs of tolerance," can therefore be questioned in the case of chironomids. The reduced larval growth rate was most likely due to zinc shortage and consequently was an indirect effect of the tolerance mechanism instead of a direct consequence of the extra energy invested in maintaining a tolerance mechanism.

It therefore appeared that metal-tolerant *C. riparius* larvae were capable of maintaining high growth rates when exposed to cadmium, even as larval mortality remained high. Increased tolerance was based on a higher excretion efficiency from the midgut epithelium, possibly due to an increased synthesis of metallothioneins and an increased efficiency of metal storage in the midgut. A direct selection on certain life-history characteristics due to metal pollution was not found, but results demonstrated reduced fitness under clean conditions, partly due to an increased need for essential metals and partly due to other genetically determined factors related to tolerance development.

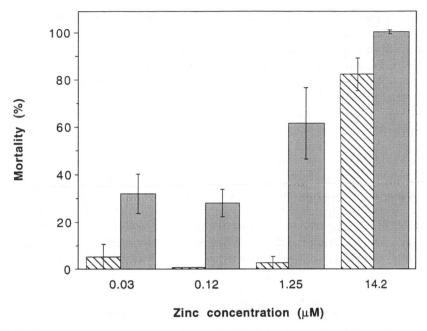

**FIG. 5.** Effects of zinc exposure on larval mortality (%) of a cadmium-tolerant, laboratory-reared population (filled bars) and a nontolerant laboratory population (hatched bars) of *C. riparius*. Mean values together with standard errors (±1 SE) are presented.

## HERITABILITY OF METAL TOLERANCE

Because differences in metal tolerance among field populations of chironomids were demonstrated using larvae from a laboratory-reared $F_1$-generation (Fig. 3; Postma et al. 1995b) the presence of a genetic component of the tolerance was assumed. However, the presence of maternal effects could not be ruled out completely. Further proof that the increased metal tolerance was at least partly caused by a genetic component could not be obtained using quantitative genetics based on parent-offspring relations because these analyses require that offspring be matched to individual parents. It is therefore necessary that a single couple reproduce successfully. For *C. riparius* this is not the case because swarms need to be formed in the twilight, from which paired insects drop out. Crossbreeding of midge populations, on the other hand, is possible. For this purpose emergence traps were designed in which adult, freshly emerged midges were caught separately in small plastic tubes that were placed just above the water surface. These plastic tubes were handled daily and tubes in which both male and female midges were present were discarded for further use. All individually caught male and female midges were placed in two different flight cages: newly emerged males from population 1 together with newly emerged females from population 2, and vice versa. The populations studied so far include the upstream reference site, DEG, and the polluted downstream location, Neerpelt; individuals from these populations most likely interbreed under field conditions (Groenendijk et al. 1998a). It is expected that for this experimental set up, indications of possible maternal effects in the Neerpelt population or sex-linked inheritance of metal adaptation can be detected by determining whether the observed metal tolerance is significantly influenced by the crossbreeding scheme. Egg masses resulting from crossbreeding of these two populations were used in short-term experiments, starting with first generation, first instar larvae. These larvae were exposed for 96 hours to a range of cadmium concentrations, after which larval growth was measured. Results clearly demonstrated that cadmium reduced the larval growth of the reference population, DEG (Fig. 6). The Neerpelt larvae, on the other hand, showed maximum growth at somewhat increased cadmium concentrations and were less affected by cadmium than DEG-larvae, even at the highest cadmium concentrations. Results of the two crossbred populations were almost identical to one another and showed intermediate responses to cadmium, as compared to growth patterns in the two parent populations, DEG and Neerpelt. Although with this experimental design it is not possible to estimate the heritability ($h^2$) of characters involved in the adaptation to metals, results showed no clear indications

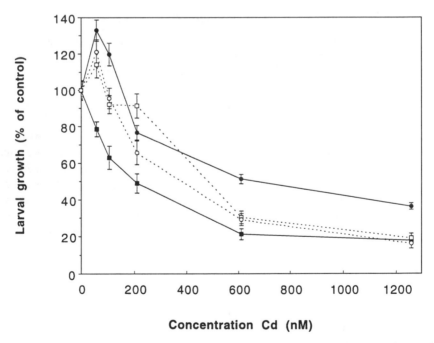

**FIG. 6.** Larval growth (as a percentage of the corresponding control) of first instar larvae exposed to cadmium. $F_1$-generation larvae were obtained from both the unpolluted upstream site DEG (filled squares) and the polluted downstream site Neerpelt (filled circles). In addition, both populations were interbred, resulting in two crosses, males DEG * female Neerpelt (open squares) and females DEG * males Neerpelt (open circles). Mean values together with standard errors (±1 SE) are presented.

of maternal effects or sex-linked inheritance, suggesting a major genetic component for the increased metal tolerance in the exposed population Neerpelt. So far, estimations of the heritability of characters involved in metal tolerance in invertebrates are only available for the springtail *Orchesella cincta* ($0.33 < h^2 < 0.48$, Posthuma et al. 1993a) and the oligochaete *Limnodrilus hoffmeisteri* ($h^2 > 0.9$, Klerks and Levinton 1993), values that indicate that adaptation to metals can develop rapidly.

Because the studied characteristics only showed relatively small differences between adapted and unadapted populations, adaptation to metals in the midge *C. riparius* seemed to be a gradual process. The same was concluded for other invertebrates such as the collembola *O. cincta* (Posthuma and van Straalen 1993). It is therefore hypothesized that in these cases the metal tolerance of an individual is determined by several partly additive factors and probably involves a polygenic response (cf. Shaw, chapter 2). Consequently, differences in the level of tolerance may arise between populations depending on, among other things, the genetic variation present for each of the additive factors. Furthermore, although the net process is gradual, a discrete and sudden increase in one of the factors involved is still possible, as is illustrated by the duplication of the metallothionein gene in *D. melanogaster* (Maroni et al. 1987). Pesticide resistance is also often the result of a polygenic response, although selection for insecticide resistance using concentrations exceeding the $LC_{100}$ of susceptible populations occasionally results in single gene responses and a large change in the resistance level within a population (McKenzie and Batterham 1994). In general, the relationship between the actual metal concentration during the selection processes and $LC_{50}$ values seems to be of vital importance for the occurrence of polygenic or single gene responses for metal tolerance.

Based on this reasoning and the actual metal concentrations in the Dommel, additional evidence for a genetic component in metal adaptation in the midge *C. riparius* was obtained by performing a multigeneration selection experiment for cadmium-tolerance in the laboratory (Postma and Davids 1995), during which midges were exposed to environmentally realistic cadmium concentrations. The effects on mortality, growth, and reproduction were studied, employing the methods of the chronic toxicity experiments described above. In addition, acute tests were performed at the end of the multigeneration experiment to further establish whether tolerance increased over the generations. Successive generations were started by using larvae originating from at least 10 egg masses. Each generation was maintained until all surviving larvae had emerged. This took between 4 and 8 weeks, depending on the cadmium concentration. The total experiment was conducted over nine consecutive generations and lasted about one year (see Postma and Davids 1995 for further details). In Fig. 7a the integrated effects of cadmium on all life-history parameters studied (mortality, growth rate, and reproduction) are presented, using the population growth rate (calculated for all replications separately). For all three treatments, these results demonstrated that the effects of cadmium significantly increased over the generations. The population exposed to 54.2 nM Cd, however, recovered, and the population growth rate of the ninth generation was significantly higher than that of the fifth to seventh generations. The population growth rate of the population exposed to 159.6 nM Cd peaked in the second

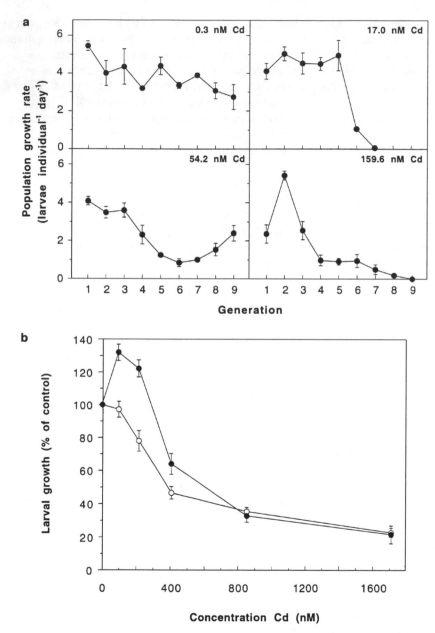

**FIG. 7.** (*a*) Population growth rates (larvae individual$^{-1}$ day$^{-1}$) during nine consecutive generations of *C. riparius* exposed to four different cadmium concentrations (upper right corner in nM). (*b*) Growth reduction by cadmium in first instar larvae of *C. riparius*. Data are expressed as percentage reduction compared to the control. Open circles represent the ninth generation of the control population and closed circles the ninth generation of the population exposed to 54.2 nM cadmium. Mean values together with standard errors (±1 SE) are presented (reproduced with permission from Postma and Davids 1995).

generation (with a value of 5.4 larvae individual$^{-1}$ day$^{-1}$, mainly due to a low mortality) but later it decreased continuously until all replicates were extinct by the ninth generation. The population exposed to 17 nM cadmium also went extinct, indicating that adaptation does not always occur. Based on these results, additional short-term experiments were conducted in which the effects of cadmium on the growth of unexposed larvae (obtained from the control population) were compared with the growth of larvae obtained from the ninth generation of the population that had been continuously exposed to 54.2 nM cadmium (Fig. 7b; based on Postma and Davids 1995).

Based on these chronic and short-term experiments it appeared that life-history characters strongly correlated with both the cadmium concentration and generation and confirmed that adaptation to metals can occur within a few generations. Furthermore, it was concluded that the increased tolerance to cadmium in the midge populations studied was mainly due to genetic factors.

## INFLUENCE OF GENE FLOW

The fourth criterion refers to the dynamic interaction between the selective pressure by elevated metal concentrations and gene flow as determined by two spatial components, pollution and population heterogeneity (Brandon 1990). Due to this interaction the presence or absence of a certain species in a contaminated habitat is not only influenced by its sensitivity or ability to adapt, but also by the rate of immigration from unpolluted sites. Gene flow can reduce the speed of adaptation to pollutants (Comins 1977; Roush and McKenzie 1987; Taylor and Georghiou 1979) but on the other hand it can, for example, reduce the effect of inbreeding and can introduce new genes that are essential to further increases in tolerance (Slatkin 1987).

A possible influence of gene flow in our study area was indicated by the differences found between the two metal adapted midge populations at Neerpelt and Borkel. For instance, the Neerpelt population, sampled near to the zinc factory, was characterized by a high control mortality, while larval development time did not differ from control populations (Fig. 3; based on Postma et al. 1995b). The Borkel population, on the other hand, was sampled 7 km further downstream and was characterized by a low control mortality but a reduced larval growth rate. It was assumed that one of the main factors responsible for these differences was the influence of gene flow. Drifting, nontolerant larvae originating from sites upstream from the point source of cadmium could easily reach the Neerpelt population, whereas the effect on the Borkel population will be much smaller. Therefore, a detailed field study was carried out to examine larval drift into the Neerpelt population, and population dynamics of reference and metal adapted midge populations were simultaneously studied (see Groenendijk et al. 1998a for details). The drift measurements were carried out at the reference location DEG, just upstream from the input from the zinc factory. As a consequence, all drifting larvae originated from non-exposed, upstream midge populations and would, after passing the point source of metal pollution, directly enter the metal-exposed site, Neerpelt. A 250 $\mu$m net (30 cm wide) was held in the middle of the river just under the water surface

GENETICS AND ECOTOXICOLOGY

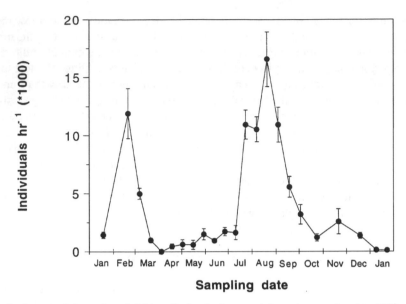

**FIG. 8.** Seasonal dynamics of drifting *C. riparius* larvae at the reference location DEG during 1995. Mean values together with standard errors (±1 SE) are presented.

for three successive 4–8 minute periods. These drift measurements demonstrated a steady transport of nontolerant larvae towards the polluted Neerpelt site, with peaks in late winter and during summer (Fig. 8; based on Groenendijk et al. 1998a). Based on these drift measurements and the actual larval densities at Neerpelt, estimates were made of the percentages of upstream larvae entering the polluted Neerpelt site per day. Assuming that larvae drift only 100 m, which is the maximum recorded distance for drifting chironomids and other invertebrates (McLay 1970; Elliott 1971; Larkin and McKone 1985; Brittain and Eikeland 1988), the estimated daily input of nontolerant larvae to the Neerpelt population regularly exceeded 100% of the local midge population present, especially in early spring (Groenendijk et al. 1998a). A preliminary study using allozyme polymorphisms further demonstrated a complete lack of population substructuring between two unexposed and two metal-exposed *C. riparius* populations in the Dommel, again indicating a high amount of gene flow (Raijmann and van Grootveld 1997). Additional experiments on morphological characters of the same populations also indicated that drifting nonadapted larvae can easily reach the polluted sites just downstream from the Eindergatloop and might even reach the Borkel population (Groenendijk et al. 1998b).

Because of the similarity in population dynamics of midges at the reference site, DEG, and Neerpelt in the field study (Groenendijk et al. 1998a), it was concluded that drifting, nontolerant larvae probably contribute to the Neerpelt population. It was therefore hypothesized that drifting, nonadapted midges can interbreed with the adapted midges from the Neerpelt population. This would decrease not only the level of metal adaptation, but also the genetic stability of the adapted population because sexual reproduction can produce offspring with a lowered fitness (due to an 'unfavorable' combination of genes). This can increase mortality rate, but the surviving larvae would be likely to have an increased metal tolerance (as was observed for the Neerpelt population grown under control conditions; see Fig. 3, Postma et al. 1995b).

It seems that the influence of gene flow can be quite high, affecting both population dynamics and cadmium-tolerance considerably. Furthermore, this influence will vary both in time and space and depend on several factors, such as population dynamics and current velocity. The resulting highly dynamic situation is in sharp contrast to other examples of metal adaptation in (terrestrial) invertebrates, in which the selective pressure is rather constant due to a more constant level of contamination or a low amount of gene flow, which is due to a comparatively low dispersal rate (cf. Posthuma 1992).

## PHYLOGENETIC INFORMATION TO ENABLE DISTINCTION BETWEEN THE ORIGINAL AND THE DERIVED STATE OF THE CHARACTER

Like evidence that selection has occurred (criterium 1), obtaining evidence about the original state of characters involved in metal adaptation is often difficult, especially in situations where the actual exposure is rather constant. In the River Dommel, studying the actual evolution of tolerance is, however, possible. As mentioned before, the selective pressure will depend on the larval-stage present, the actual metal concentration in the water, and on gene flow from sites upstream. Fluctuations in factors such as population dynamics, rainfall, and discharges might therefore directly influence the level of tolerance present in the field populations. Adaptation to metals in river-inhabiting chironomids should therefore be looked upon as a dynamic situation in which the actual level of tolerance gradually fluctuates and is perhaps sometimes very low or even absent. In these cases, the original state of a character involved in tolerance might best be described by the state of this character at times in which the level of metal adaptation is low.

Support for such a dynamic situation in the actual level of metal adaptation was obtained in a recent study in which seasonal measurements were carried out on two field populations: the reference location DEG, just upstream from the zinc factory, and the polluted Neerpelt location, situated 500 m downstream. Short-term $EC_{50}$ values (96 hours, measured as larval growth) for cadmium were determined for unexposed, laboratory-reared first generation larvae, following the methods for short-term experiments described above (section 3, heritability). $EC_{50}$ values for the reference population DEG were stable between August and December 1996 ($\pm200-250$ nM Cd), whereas $EC_{50}$ values for the Neerpelt population fluctuated between $<200$ and $>700$ nM cadmium (Fig. 9). Furthermore, control mortality also fluctuated considerably in the Neerpelt population when compared to the reference site. It is therefore concluded that the level of adaptation to metals in the Neerpelt midge population is subject to strong fluctuations and can sometimes even be absent, which again indicates that "nonadapted" is most likely the original state.

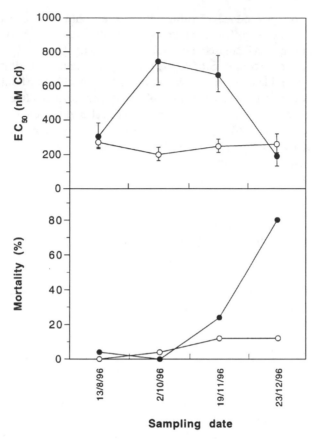

**FIG. 9.** Seasonal measurements of short-term (96 hours) $EC_{50}$-values for larval growth (upper graph) and control mortality (lower graph) for the reference location DEG (open symbols) and the polluted downstream site Neerpelt (black symbols). $EC_{50}$ values were calculated by fitting a model for logistic response. Error bars represent 95% confidence limits.

## CONCLUSIONS

This is the first set of studies providing conclusive evidence of genetic adaptation to metals in aquatic insects, according to Brandon's criteria (1990). More important, however, are the recent findings on larval drift of chironomids and the influence of the resulting gene flow on the level of adaptation because these are among the first detailed experiments on the subject concerning invertebrates. Our results demonstrated that the actual level of metal tolerance in midge populations fluctuates strongly, varying in both time and space. This dynamic situation is influenced by several factors, such as population dynamics and current velocity, and the interaction between such factors might even cause a temporary absence of tolerance.

## ACKNOWLEDGMENTS

The authors would like to thank Marion Buckert-de Jong, Chris de Groot, and many M.Sc. students for their help in sampling and analysis, Wim Admiraal and Michiel Kraak for their comments, and Heather Leslie for correcting the English. The investigations were partly funded by the Netherlands Foundation for Life Sciences (SLW), which is subsidised by the Netherlands Organization for Scientific Research (NWO).

## REFERENCES

Bajraktari, I., G. Savic, D. Marinkovic, and A. Hajrizi. 1987. The influence of heavy metals on the duration of *Drosophila melanogaster* preadult development. *Acta Biologiae et Medicinae Experimentalis* 12:57–66.

Brandon, R. N. 1990. *Adaptation and environment*. Princeton, New Jersey: Princeton University Press.

Brittain, J. E., and T. J. Eikeland. 1988. Invertebrate drift—a review. *Hydrobiologia* 166:77–93.

Brown, B. E. 1976. Observations on the tolerance of the isopod *Asellus meridianus* Racovitza 1919 to copper and lead. *Water Research* 10:555–559.

Charlesworth, B. 1980. *Evolution in age-structured populations*. Cambridge: Cambridge University Press.

Comins, H. N. 1977. The development of insecticide resistance in the presence of migration. *Journal of Theoretical Biology* 64:177–197.

Cook, S. C. A., C. Lefébre, and T. McNeilly. 1972. Competition between metal tolerant and normal plant populations on normal soil. *Evolution* 26:366–372.

Cox, R. M., and T. C. Hutchinson. 1981. Multiple and co-tolerance to metals in the grass *Deschampsia cespitosa*: Adaptation, preadaptation and 'costs'. *Journal of Plant Nutrition* 3:731–741.

Dallinger, R. 1993. Strategies of metal detoxification in terrestrial invertebrates. In *Ecotoxicology of metals in invertebrates*, eds. R. Dallinger and P. S. Rainbow, 245–289. Boca Raton, FL: Lewis Publishers.

Davies, B. R. 1976. Wind distribution of the egg masses of *Chironomus anthracinus* (Zetterstedt) (Diptera: Chironomidae) in a shallow, wind-exposed lake (Loch Leven, Kinross). *Freshwater Biology* 6:421–424.

Donker, M. H., C. Zonneveld, and N. M. van Straalen. 1993. Early reproduction and increased reproductive allocation in metal adapted populations of the terrestrial isopod *Porcellio scaber*. *Oecologia* 96:316–323.

Elliott, J. M. 1971. The distances travelled by drifting invertebrates in a Lake District stream. *Oecologia* 6:350–379.

Falconer, D. S. 1981. *Introduction to quantitative genetics*. New York: Longman.

Gadgil, M., and W. H. Bossert. 1970. Life historical consequences of natural selection. *American Naturalist* 104:1–24.

Groenendijk, D., J. F. Postma, M. H. S. Kraak, and W. Admiraal. 1998a. Seasonal dynamics and larval drift of *Chironomus riparius* (Diptera) populations in a metal contaminated lowland river. *Aquatic Ecology*, submitted.

Groenendijk, D., L. Zeinstra, and J. F. Postma. 1998b. Fluctuating asymmetry and mentum gaps in populations of the midge *Chironomus riparius* (Diptera, Chironomidae) from a metal contaminated river. *Environmental Contamination and Toxicology*, 17(10): (in press).

Hickey, D. A., and T. McNeilly. 1975. Competition between metal tolerant and normal plant populations: A field experiment on normal soil. *Evolution* 29:458–464.

Holloway, G. J., R. M. Sibly, and S. R. Povey. 1990. Evolution in toxin-stressed environments. *Functional Ecology* 4:289–294.

Klerks, P. L., and J. S. Levinton. 1989. Rapid evolution of metal resistance in a benthic oligochaete inhabiting a metal-polluted site. *Biological Bulletin* 176:135–141.

Klerks, P. L., and J. S. Levinton. 1993. Evolution of resistance and changes in community composition in metal-polluted environments: A case study on Foundry Cove. In *Ecotoxicology of metals in invertebrates*, eds. D. Dallinger and P. S. Rainbow, 223–241. Boca Raton, FL: Lewis Publishers.

Klerks, P. L., and J. S. Weis. 1987. Genetic adaptation to heavy metals in aquatic organisms: A review. *Environmental Pollution* 45:173–205.

Larkin, P. A., and D. W. McKone. 1985. An evaluation by field experiments of the McLay model of stream drift. *Canadian Journal of Fisheries and Aquatic Sciences* 42:909–918.

Lauverjat, S., C. Ballan-Dufrancais, and M. Wegnez. 1989. Detoxification of cadmium, ultrastructural study and electron-probe microanalysis of the midgut in a cadmium resistant strain of *Drosophila melanogaster. Biology of Metals* 2:97–107.

Maroni, G., J. Wise, J. E. Young, and E. Otto. 1987. Metallothionein gene duplications and metal tolerance in natural populations of *Drosophila melanogaster. Genetics* 117:739–744.

McKenzie, J. A., and P. Batterham. 1994. The genetic, molecular and phenotypic consequences of selection for insecticide resistance. *Trends in Ecology and Evolution* 9:166–169.

McLay, C. 1970. A theory concerning the distance travelled by animals entering the drift of a stream. *Journal of the Fisheries Research Board of Canada* 27:359–370.

Miller, M. P., and A. C. Hendricks. 1996. Zinc resistance in *Chironomus riparius*: Evidence for physiological and genetic components. *Journal of the North American Benthological Society* 15:106–116.

Mulvey, M., and S. A. Diamond. 1991. Genetic factors and tolerance acquisition in populations exposed to metals and metalloids. In *Metal ecotoxicology: Concepts and applications*, eds. M. C. Newman and A. W. McIntosh. 301–321. Boca Raton: Lewis Publishers.

Nevo, E., R. Ben-Shlomo, and B. Lavie. 1984. Mercury selection of allozymes in marine organisms: Prediction and verification in nature. *Proceedings of the National Academy of Science USA* 81:1258–1259.

Pascoe, D., K. A. Williams, and D. W. J. Green. 1989. Chronic toxicity of cadmium to *Chironomus riparius*—Effects on larval development and adult emergence. *Hydrobiologia* 175:109–115.

Posthuma, L. 1992. Genetic ecology of metal tolerance in Collembola. Ph.D. Thesis, Amsterdam, The Netherlands: Free University.

Posthuma, L., R. F. Hogervorst, and N. M. van Straalen. 1992. Adaptation to soil pollution by cadmium excretion in natural populations of *Orchesella cincta* (L.) (Collembola). *Archives of Environmental Contamination and Toxicology* 22:146–156.

Posthuma, L., R. F. Hogervorst, E. N. G. Joosse, and N. M. van Straalen. 1993a. Genetic variation and covariation for characteristics associated with cadmium tolerance in natural populations of the springtail *Orchesella cincta* (L.). *Evolution* 47:619–631.

Posthuma, L., R. A. Verweij, B. Widianarko, and C. Zonneveld. 1993b. Life-history patterns in metal adapted Collembola. *Oikos* 67:235–249.

Posthuma, L., and N. M. van Straalen. 1993. Heavy-metal adaptation in terrestrial invertebrates: A review of occurrence, genetics, physiology and ecological consequences. *Comparative Biochemistry and Physiology* 106C:11–38.

Postma, J. F., M. C. Buckert-de Jong, N. Staats, and C. Davids. 1994. Chronic toxicity of cadmium to *Chironomus riparius* (Diptera: Chironomidae) at different food levels. *Archives of Environmental Contamination and Toxicology* 26:143–148.

Postma, J. F., and C. Davids. 1995. Tolerance induction and life-cycle changes in cadmium exposed *Chironomus riparius* (Diptera) during consecutive generations. *Ecotoxicology and Environmental Safety* 30:195–202.

Postma, J. F., S. Mol, H. Larsen, and W. Admiraal. 1995a. Life-cycle changes and zinc shortage in cadmium tolerant midges, *Chironomus riparius* (Diptera), reared in the absence of cadmium. *Environmental Toxicology and Chemistry* 14:117–121.

Postma, J. F., A. van Kleunen, and W. Admiraal. 1995b. Alterations in life-history traits of *Chironomus riparius* (Diptera) obtained from metal contaminated rivers. *Archives of Environmental Contamination and Toxicology* 29:469–475.

Postma, J. F., P. van Nugteren, and M. C. Buckert-de Jong. 1996. Increased cadmium excretion in metal adapted populations of the midge *Chironomus riparius* (Diptera). *Environmental Toxicology and Chemistry* 15:332–339.

Raijmann, L. E. L. and I. van Grootveld. 1997. A pilot study of the use of allozyme polymorphisms to identify metal adapted populations of the midge *Chironomus riparius* Meigen. *Proceedings of the Experimental and Applied Entomology Section of the Netherlands Entomological Society (N.E.V.)*, Amsterdam, 8:93–98.

Roush, R. T., and J. A. McKenzie. 1987. Ecological genetics of insecticide and acaricide resistance. *Annual Review of Entomology* 32:361–380.

Seidman, L. A., G. Bergtrom, D. J. Gingrich, and C. C. Remsen. 1986a. Accumulation of cadmium by the fourth instar larva of the fly *Chironomus thummi. Tissue and Cell* 18:395–405.

Seidman, L. A., G. Bergtrom, and C. C. Remsen. 1986b. Structure of the larval midgut of the fly *Chironomus thummi* and its relationship to sites of cadmium sequestration. *Tissue and Cell* 18: 407–418.

Sibly, R. M., and P. Calow. 1989. A life-cycle theory of responses to stress. *Biological Journal of the Linnean Society* 37: 101–116.

Slatkin, M. 1987. Gene flow and the geographical structure of natural populations. *Science* 236: 787–792.

Sohal, R. S., and R. E. Lamb. 1977. Intracellular deposition of metals in the midgut of the adult housefly, *Musca domestica. Journal of Insect Physiology* 23:1349–1354.

Stuijfzand, S. C., D. Shen, M. Helms, and M. H. S. Kraak. 1996. Effects of pollution in the River Meuse on the midge *Chironomus riparius. Proceedings of the Experimental and Applied Entomology Section of the Netherlands Entomology Society (N.E.V.)*, Amsterdam, 7:211–216.

Taylor, C. E., and G. P. Georghiou. 1979. Suppression of insecticide resistance by alteration of gene dominance and migration. *Journal of Economic Entomology* 72:105–109.

Timmermans, K. R., W. Peeters, and M. Tonkes. 1992. Cadmium, zinc, lead, and copper in *Chironomus riparius* (Meigen) larvae (Diptera, Chironomidae): Uptake and effects. *Hydrobiologia* 241:119–134.

van Hattum, B., K. R. Timmermans, and H. A. Govers. 1991. Abiotic and biotic factors influencing in situ trace metal levels in macroinvertebrates in freshwater ecosystems. *Environmental Toxicology and Chemistry* 10:275–292.

van Straalen, N. M., T. B. A. Burghouts, M. J. Doornhof, G. M. Groot, M. P. M. Janssen, E. N. G. Joosse, J. H. van Meerendonk, J. P. J. J. Theeuwen, H. A. Verhoef, and H. R. Zoomer. 1987. Efficiency of lead and cadmium excretion in populations of *Orchesella cincta* (Collembola) from various contaminated forest soils. *Journal of Applied Ecology* 24:953–968.

Wentsel, R., A. McIntosh, and G. Atchison. 1978. Evidence of resistance to metals in larvae of the midge *Chironomus tentans* in a metal contaminated lake. *Bulletin of Environmental Contamination and Toxicology* 20:451–455.

Yamamura, M., K. T. Suzuki, S. Hatakeyama, and K. Kubota. 1983. Tolerance to cadmium and cadmium-binding proteins induced in the midge larva, *Chironomus yoshimatsui* (Diptera, Chironomidae). *Comparative Biochemistry and Physiology* 75C:21–24.

Genetics and Ecotoxicology
Edited by V. E. Forbes
Copyright © 1999 Taylor & Francis

# 6

# The Influence of Contamination Complexity on Adaptation to Environmental Contaminants

## Paul L. Klerks

**Abstract.** Many polluted areas contain complex mixtures of contaminants, motivating an assessment of the influence of this complexity on the development of contaminant resistance. The first of two approaches used here is an evaluation of theoretical predictions. An increase in the number of contaminants may have various effects on genetic diversity (ranging from a decrease in diversity as a consequence of reductions in population size and selective removal of sensitive phenotypes to a possible increase in diversity due to an increased mutation rate). It is further expected that an increase in the number of contaminants will (at a combined toxicant effect comparable to that for an individual contaminant) reduce the selection intensity for the individual contaminants. Moreover, theoretical considerations predict that the incidence of negative genetic correlations among traits will increase as selection proceeds, thereby further reducing the effectiveness of selection for resistance to a mixture of contaminants. The second approach used here is an investigation into the occurrence of contaminant adaptation in several species of aquatic organisms inhabiting sites in Louisiana contaminated by a large number of different contaminants. Mosquitofish (*Gambusia affinis*) in Bayou Trepagnier show a reduced sensitivity to lead but not zinc. However, the lead resistance seems to be due to physiological acclimation rather than adaptation. In Pass Fourchon on the Gulf of Mexico, none of the three species studied (the darter goby *Gobionellus boleosoma*, the polychaete *Streblospio benedicti* and the grass shrimp *Palaemonetes pugio*) shows evidence of adaptation to the toxicity of sediment contaminated by produced water discharged there in four decades, despite strong evidence of sediment toxicity. Although it is possible that factors other than contamination complexity are responsible for the lack of adaptation in these four species, this result is consistent with the general prediction from the theoretical considerations. It also agrees with pesticide resistance studies that generally indicate a reduced rate of resistance development when several independently acting pesticides are used in combination. Although further research is needed on this issue, current understanding indicates that it may be unlikely that adaptations can ensure the long-term survival of most populations inhabiting areas polluted by a variety of different contaminants.

**Keywords.** Genetic diversity; metals; multiple contaminants; PAHs; resistance; selection intensity.

## INTRODUCTION

It has been well established that populations exposed to a contaminant may develop a genetically-based resistance to the chemical. The evolution of resistance is best known for pesticide resistance in target species. At least 504 species of insects and mites were resistant to pesticides as of 1988 (Georghiou 1990). Adaptations have also been reported in populations exposed to environmental contaminants. In addition to the present volume, several reviews have been published on this topic (including Rahel 1981; Klerks and Weis 1987; Klerks 1989; Posthuma and Van Straalen 1993). The literature indicates that such adaptations do occur, though the likelihood of these occurrences remains to be determined. This conclusion is not unique to the field of contaminant adaptation as similar uncertainties exist for adaptations to global environmental change (Travis and Futuyma 1993).

It is not clear if it is less likely that resistance will develop in populations exposed to multiple contaminants. As a general rule, adaptations to environmental contaminants are best known for populations inhabiting environments contaminated by one major toxicant, or a small number of toxicant. Examples include adaptations to metals in plants on abandoned mines (Shaw, chapter 2), a polychaete inhabiting rivers draining these areas (Bryan and Hummerstone 1971, 1973), an oligochaete inhabiting a marsh contaminated by discharges from a facility producing Cd-Ni batteries (Klerks and Levinton 1989b), in a chironomid inhabiting metal-contaminated river sites (Postma and Groenendijk, chapter 5), and a collembola and an isopod inhabiting metal mine and smelter sites (Donker and Bogert 1991; Posthuma et al. 1992). Similarly, selection experiments that investigate whether laboratory populations will adapt to contamination have only been conducted with exposures to one or a few contaminants (Rahel 1981; LeBlanc 1982; Velázquez et al. 1987; Klerks and Levinton 1989a,b; Diamond et al. 1995). However, many polluted environments contain a large number of different contaminants, typically both metals and various organic contaminants. Thus, it is important to investigate whether the occurrence of adaptations to environmental contamination is affected by contaminant complexity.

This chapter takes two approaches. First, theoretical considerations are brought forward that predict changes in the likelihood of adaptations when the number of contaminants is increased. Second, results are presented from my own investigations into the occurrence of contaminant adaptation in several species of aquatic organisms inhabiting sites contaminated by a large number of different toxicants. The theoretical considerations are largely based on the population genetics of quantitative characters, where complex equations and terminology have been avoided because many ecotoxicologists may be unfamiliar with this field. Although my evaluations may thus be somewhat oversimplified, I nevertheless feel that our insight into adaptation to environmental contaminants (especially as this issue becomes more complicated and realistic when factors such as contamination complexity are included) benefits from this theoretical framework.

## THEORETICAL CONSIDERATIONS

Various models have been developed for predicting the response to selection for a quantitative trait such as resistance. In the basic model, the response to selection is solely determined by the amounts and relative proportions of additive genetic variation $(V_A)$ and phenotypic variation $(V_P)$ for the trait in the population and by the intensity of selection (Falconer 1981; Roff 1997). The ratio $V_A/V_P$ denotes the heritability. The response to selection, is directly proportional to the heritability, the intensity of selection, and $\sqrt{V_P}$ of the trait (Falconer 1981). The above model assumes truncation selection, in contrast to some other models that use Gaussian selection with a moving optimum (e.g., Lynch et al. 1991; Lynch and Lande 1993). The latter model may be more relevant to responses to an environmental change such as an increased temperature (for which there is clearly an optimum value), but contaminants are more likely to cause a truncation selection with only the most resistant phenotypes contributing to the next generation. Advanced models also include factors such as random genetic drift and linkage disequilibrium (Bulmer 1989), population size and the number of contributing loci (Roff 1997), and the input of additional variation by mutation (Hill 1982; Lynch et al. 1991). These factors are not included in the basic model. For the sake of clarity, the following population genetic considerations will rely most heavily on the basic model of the response to selection while incorporating some predictions obtained from the advanced models.

The first determinant of selection response considered here is the amount of genetic variation. Environmental contamination may affect the amount of genetic variation in a population through several processes. Contaminant toxicity will result in a reduced population size. There is evidence that strong reductions in population size can significantly reduce the amount of genetic variation in a population (Nei et al. 1975; Leberg 1992; cf. Snell et al. chapter 9), and rare alleles are especially likely to be lost. This argument is supported by empirical evidence that the response to selection is more rapid in large populations than in small populations (Simberloff 1988; Weber and Diggins 1990). The reduced amount of genetic variation in a small population appears to be not only due to the lower initial availability of genetic variation compared to that in a larger population, but also to the reduction in the accumulation of new mutations in a small population (Weber and Diggins 1990). Moreover, the selection exerted by a contaminant will cause a further reduction in the amount of genetic variation for resistance to that contaminant by eliminating the more sensitive phenotypes from a population. Several studies have looked at genetic variation in populations inhabiting contaminated environments, with most reporting a decrease in genetic variation relative to populations from a pristine site (Kopp et al. 1992; Benton et al. 1994; Murdoch and Hebert 1994) though some others did not detect such differences (Nevo et al. 1978; Frati et al. 1992). An increase in contaminant complexity is thus expected to bring about an increased loss of genetic variation, as population sizes would be further reduced and more phenotypes would be eliminated (the ones sensitive to any of the contaminants or their interactions) relative to the situation with one contaminant present. An observation of a lower response to selection in a central population than that in a population at the margin of a species range (Hoffmann and Blows 1993) agrees with this predicted effect of an increase in environmental stress. The arguments brought forward so far indicate a reduction in the amount of genetic variation (and thus a reduction in the response to selection) as the number of contaminants increases. However, other factors may have the opposite effect. There is some evidence that the amount of additive genetic variation (as well as phenotypic variation) for a trait may increase as contamination increases (Holloway et al. 1990; Forbes 1996). Similarly, it has been suggested that mutational input to genetic variation is positively correlated with environmental stress (Hoffmann and Parsons 1991). In addition, the genetic variance for a trait is expected to be higher for traits affected by more loci (Houle 1991), which leads to the prediction that there will be more genetic variation for resistance to a combination of contaminants (likely to be affected by more genes) than for resistance to a single contaminant. Further insight into the magnitude of the two groups of opposing forces (increases and decreases in the amount of genetic variation) is needed for a more definitive assessment of the net effect on genetic variation. A review of studies that investigated the effect of an increase in stress on the heritability of a trait (reviewed by Hoffmann and Parsons 1991) indicated that the net effect may be dependent on specific conditions, including increases, decreases, and absences of changes in heritability.

Another determinant of selection response that is evaluated here for its relationship to contaminant complexity is the intensity of selection. The intensity of selection for a trait is determined by the values for the trait in the selected individuals (those that contribute to the next generation) relative to the population mean. The selection intensity will be higher if a more select group of individuals (with respect to the trait under selection) contributes to the next generation. Thus when one contaminant is present at detrimental levels, the selection intensity will tend to increase as a function of the severity of the contamination. However, with several contaminants acting simultaneously and independently, the resulting increase in mortality (relative to the situation of only one contaminant present) does not result in an increased selection pressure. To give a simple numerical example, let us assume that two specific contaminants each result in a 70% mortality when present alone and a 91% mortality when present together (survivors to both insults would be 30% of 30% and thus be 9%). These 9% belong on average to the top 30% with respect to resistance to each of the two contaminants. Thus in spite of the 91% mortality, the intensity of selection for resistance is equivalent to a 70% mortality. This difference becomes more pronounced with more contaminants (e.g., the same intensity of selection at a total mortality of 99.8% when five contaminants are involved). Another way to look at this is to compare the intensities of selection at a specific mortality rate, e.g., 90%. With one contaminant present, this mortality would result in an intensity of selection of 1.755 (Falconer 1981). If this same mortality is due to more independently acting contaminants, the intensity of selection decreases with an increase in the number of contaminants (Fig. 1). Because the response to selection is directly and linearly dependent on the intensity of selection, the increase in resistance would be substantially

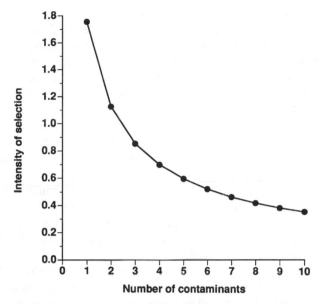

**FIG. 1.** Intensity of selection as a function of the number of independently-acting contaminants when only those individuals that are in the top 10% (with respect to resistance to these contaminants) contribute to the next generation. This is a hypothetical example.

less as the number of contaminants responsible for the mortality increases (e.g., the selection response would decrease by about two-thirds if the number of contaminants exerting the selection is increased from one to five).

The argument of a relative decrease in the intensity of selection exerted by the contamination as a function of the number of contaminants involved assumes an independent action of the contaminants (and thus also an absence of cross-resistance once resistance evolves). However, this may not be the case. For example, two contaminants may have a shared mode of action. Absence of independence can also be considered at the genetic level; there may be a genetic correlation for resistance to the two contaminants. This could be due for example to genes for the two different traits being in close proximity on a chromosome, or it could be the consequence of one gene affecting both traits. If a genetic correlation for resistance to two contaminants is at its theoretical maximum (1.0), the intensity of selection at a specific mortality rate may not be diminished relative to the same mortality rate brought about by one contaminant. However, correlated responses appear much less predictable than the direct response to selection (Gromko et al. 1991). Genetic correlations among stress resistance traits have been reported from a variety of studies, especially in *Drosophila* (reviewed in Hoffmann and Parsons 1991). This literature reveals strong positive correlations in some cases, whereas correlations were absent in others. It is expected that genetic correlations may be lower for chemical stresses than for others on the basis of the high specificity of underlying mechanisms of resistance to chemicals (Hoffmann and Parsons 1991). A contrasting argument could be made by arguing that contaminated environments are novel, and there is some evidence that genetic correlations may be higher in such environments (Service and Rose 1985). Although the exact values of genetic correlations cannot be predicted easily, it is safe to conclude that genetic correlations among chemical resistance traits are on average well below 1. In spite of positive genetic correlations among traits, selection for resistance to multiple traits will therefore still be significantly slower compared to the case with selection exerted by only one contaminant.

Although the presence of positive genetic correlations among traits reduces the effect of contamination complexity on the response to selection, these positive genetic correlations may be short-lived (Falconer 1981). Genes that affect multiple traits in the direction of the selection (increased resistance) should be strongly acted on by selection and rapidly brought to fixation. This would mean that the positive correlations among the traits disappear early in the selection process. Furthermore, it is hypothesized that negative genetic correlations will become more prevalent because these are less affected by selection than are positive correlations. The situation may be less predictable because the occurrence of negative correlations may be in part determined by the genes' underlying functional architecture (Houle 1991). Negative genetic correlations may halt further increases in resistance in spite of a continued selection pressure in the presence of heritable variation for the traits (Falconer 1981). Experimental evidence for the predicted increased prevalence of negative genetic correlations during the course of selection is conflicting (Sheridan and Barker 1974), though strong negative genetic correlations have been reported for resistance to some pesticides in target species (e.g., Cilek and Knapp 1993).

Physiological considerations also arrive at the conclusion that adaptations are less likely when more contaminants are involved. Adaptations require the presence of an underlying physiological or biochemical mechanism that reduces the toxic effects. Examples of such mechanisms are the detoxification of specific metals by metallothioneins and the reduction in contaminant influx through a sodium channel. Adaptations may then occur as a consequence of the contaminant exerting a selective pressure that acts on genetic variation in a population for the expression of these mechanisms. With more contaminants (and their interactions) involved, the chance that such mechanisms and variation for these mechanisms exist in a population becomes progressively smaller.

Finally, a successful adaptation requires the population to withstand the various consequences of the initial population size reductions resulting from the toxic effects. Environmental contamination could cause the population level to fall below the minimum required for maintaining the population (e.g., density too low for finding a mate). In addition, a population bottleneck may result in a significant degree of inbreeding and therefore cause inbreeding depression. Inbreeding depression has been reported for various laboratory and natural populations (e.g., Van Noordwijk and Scharloo 1981; Jarne and Delay 1990; Ribble and Millar 1992; Chen 1993; Latter and Mulley 1995). Inbreeding depression could severely affect a population's continued existence, though no empirical evidence has yet been reported confirming that inbreeding depression may result in the demise of natural populations (Caro and Laurenson 1994).

Because reductions in population size generally are expected to be more severe if the number of contaminants increases, the factors listed in this paragraph should contribute negatively to the relationship between a population's prospect for a successful adaptation and the number of contaminants in its environment.

In conclusion, although factors that bring about a decrease in genetic variation as contamination complexity increases could possibly be offset by other factors with an opposing effect, additional considerations strengthen the prediction that the response to selection will tend to decrease as the number of contaminants exerting that selection increases.

## CASE STUDIES ON RESISTANCE AT SITES WITH COMPLEX MIXTURES OF CONTAMINANTS

### Mosquitofish in Bayou Trepagnier

Bayou Trepagnier is a small Louisiana stream that flows into Bayou LaBranche just before the latter empties into Lake Pontchartrain. Bayou Trepagnier has been contaminated by waste discharges (during the periods 1920–1929 and 1951–1995) from a petrochemical facility. These discharges have resulted in contamination by chromium, lead, zinc, chloroform, dichlorobenzene, benzene, toluene, ethylbenzene, and a long list of polynuclear aromatic hydrocarbons (LADEQ 1989; DeLaune et al. 1990; Pardue et al. 1992). It has been reported that fish and benthic communities in a section of the bayou have been impacted and that water and sediment are toxic in laboratory bioassays (LADEQ 1989). Resistance to lead and zinc has been investigated in the mosquitofish *Gambusia affinis* inhabiting this site (Klerks and Lentz 1998). A nearby waterbody (Engineer's Canal) served as the control site.

Levels of zinc and lead were determined in mosquitofish at the beginning of this study to verify that these contaminants were currently bioavailable to mosquitofish. Lead concentrations in mosquitofish collected at Engineer's Canal were $138 \pm 25$ ng/g dry weight (mean $\pm$ S.E.), whereas tissue levels of those collected at Bayou Trepagnier were $898 \pm 306$ ng/g. Differences in zinc tissue levels were less pronounced, with values for the two sites of $228 \pm 13$ and $277 \pm 9$ $\mu$g/g, respectively. These differences were statistically significant ($p < 0.05$ in ANOVA on log-transformed data) for both elements.

Differences in resistance between the populations were determined three days after collection of the fish from the field sites. Resistances to Zn and Pb were quantified in fish (either sex, 19–25mm standard length) in 96 h exposures to these metals in soft reconstituted fresh water (APHA et al. 1995). For the lead bioassays, sodium chloride was substituted for sodium bicarbonate in the recipe for reconstituted water to prevent the formation of a precipitate when the $Pb(NO_3)_2$ was added. A total of five concentrations (plus control) were used for each of the exposures (10 or 15 fish per concentration, using five fish per exposure bowl with 2.5 l of water). Zinc and lead concentrations ranged from 3.9 to 30 mg/l and from 2.6 to 20 mg/l, respectively.

**TABLE 1.** *Results from bioassays comparing resistance of* Gambusia affinis *from Bayou Trepagnier (BT) and the control site Engineer's Canal (EC)*[1]

| Exposure | Statistical outcome (effect of collection site) |
|---|---|
| Exposed to Zn shortly after field-collection | BT = EC ($p = 0.795$) |
| Exposed to Pb shortly after field-collection | BT > EC ($p = 0.024$ in exp. 1) |
| | BT > EC ($p < 0.0001$ in exp. 2) |
| Exposed to Pb following deacclimation | BT = EC ($p = 0.528$) |

[1] Fish were exposed to lead or zinc three days after their collection from the field or after being kept for an additional 34 days in clean water ("deacclimation"). Resistance quantified as survival time in the exposures. Statistical results are from a survival analysis using a Weibull model testing for effects of collection site. Reprinted with permission from Klerks and Lentz (1998).

Survival was checked at 11 time points during the 96 h exposures. These data were used to compare survival times between Bayou Trepagnier and Engineer's Canal fish using survival analysis (e.g., Collett 1994) similar to the procedure used by Newman and Aplin (1992). The analyses used the parametric Weibull regression model, which included exposure concentration, exposure bowl, fish standard length, and fish sex in addition to exposure site.

For zinc, the survival analysis did not indicate a difference in survival time between the fish from Bayou Trepagnier and their conspecifics from the control site (Table 1). Thus, there is no evidence that Bayou Trepagnier mosquitofish have acclimated or adapted to zinc. This may well be a consequence of low zinc bioavailability. Although the difference in Zn tissue levels between mosquitofish from Bayou Trepagnier and their conspecifics from the control area was statistically significant, tissue levels in Bayou Trepagnier were elevated by only 21%. This small difference in tissue levels is in sharp contrast with a more than seven-fold difference in sediment zinc levels (Klerks and Lentz 1998).

Two replicate experiments investigated inter-population differences in resistance to lead. In both experiments, mosquitofish from Bayou Trepagnier survived significantly longer than fish from the control area (Table 1), demonstrating that the two populations differ in their resistance to lead. This contrast with the situation for zinc is in line with the much larger difference in lead tissue levels between the two areas (a 6.5-fold difference) than was observed for zinc. The difference between the two populations in their resistance to lead was no longer evident after fish from both sites were kept for 34 days in the laboratory in a flow-through system with dechlorinated tapwater (Table 1). The difference in lead resistance may therefore be a consequence of acclimation rather than adaptation. However, it is possible that there is a genetic difference in lead resistance between the populations in combination with a requirement for a pre-exposure for such a difference in resistance to occur (e.g., due to the induction of a detoxification mechanism). However, a nongenetic basis for the difference in resistance to lead appears to be the most parsimonious explanation for these results in the absence of further information.

## Three Species of Benthic Organisms at Pass Fourchon

Pass Fourchon is a canal in a tidal marsh area near Grand Isle, Louisiana. The research sites at Pass Fourchon are in close proximity (500–1200 m in a straight line) to the Gulf of Mexico. Produced waters (water trapped with oil and/or gas and separated from the hydrocarbons following the extraction) were discharged into the canal from 1950 to 1994. This has resulted in elevated sediment levels of several metals and a large number of polynuclear aromatic hydrocarbons (PAHs) (Rabalais et al. 1991; Klerks et al. 1997). Prominent among the PAHs are the various naphthalenes, acenaphthene, fluorene, the various dibenzothiophenes and phenanthrenes, fluoranthene, pyrene, benzanthracene, chrysene, and the benzofluoranthenes. Metal contamination is mostly limited to barium, nickel, and zinc.

Earlier research had shown that the abundance and diversity of the benthic fauna were negatively affected by the produced water discharge in Pass Fourchon and that sediment contaminant levels decrease with distance from the discharge site (Rabalais et al. 1991). I conducted a survey in 1994 to determine if effects on the benthic community were still evident at the time that the resistance study was conducted. Sediment grab samples were collected on two occasions (March and April 1994) at sites 400, 600, 800, 1200, and approximately 2500 m (the latter being the control site). Results showed that patterns in benthos abundance and diversity were similar to those obtained during 1989 and 1990 by Rabalais et al. (1991). The number of taxa exhibited a significant positive regression on the distance from the discharge area (Fig. 2a). A similar pattern was observed for the macrobenthos density (Fig. 2b), though the regression was not statistically significant (probably due to the large intra-site variation). The patterns in abundance and diversity indicate that the contaminants were continuing to have an impact on the benthic community. This is in line with PAH analyses conducted on sediment collected in Pass Fourchon in 1994. Sediment levels of PAHs at the sites were still high; total PAH levels were 15,229 ng/g dry sediment at 350 m from the discharge area (Klerks et al. 1997).

Research on resistance to the contaminants at Pass Fourchon was conducted with three benthic species: the darter goby *Gobionellus boleosoma*, the polychaete *Streblospio benedicti*, and the grass shrimp *Palaemonetes pugio*. All three species are abundant in the Pass Fourchon area and live in close contact with the contaminated sediment. *S. benedicti* is a component of the benthic infauna and a deposit feeder; *G. boleosoma* and *P. pugio* belong to the benthic epifauna. The darter goby feeds mainly on benthic meio- and macrofauna (Carle and Hastings 1982; Toepfer and Fleeger 1995), whereas the grass shrimp is a detritivore that also feeds on meiofauna and small benthic infauna (Anderson 1985).

The field survey demonstrated that the contaminated sediment had an impact on the benthic community as a whole, yet this does not necessarily mean that the species used in this project were also negatively affected by the PAH-rich sediment. Because no adaptation would be expected in the absence of a selection pressure exerted by the contamination, initial laboratory experiments assessed the potential for toxicity of the Pass Fourchon sediment in the three species. Control area darter gobies were

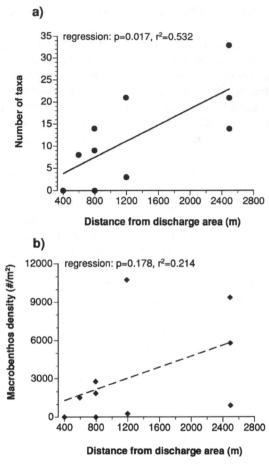

**FIG. 2.** Number of taxa (*a*) and macrobenthos density (*b*) as a function of distance from the discharge area in a macrobenthos survey of the Pass Fourchon produced water discharge area. Individual values for sediment grabs and results from regression analyses are shown.

exposed individually in 500-ml glass beakers with 100 ml Pass Fourchon or control sediment and 400 ml overlying water that was aerated slowly for a 7-week duration. Control area *S. benedicti* were exposed in groups of 10 individuals (at 3 replicates per treatment) in 1000-ml beakers with 1 cm control or Pass Fourchon sediment and 825 ml of 15 ppt. salinity water that was aerated slowly. The beaker contents were checked after 36 days for worm survival and reproduction. The exposures of the gobies and polychaetes were performed under static conditions, where distilled water was added weekly to compensate for evaporation. Grass shrimp were exposed in the larval stage with 3-day old larvae obtained from control area *P. pugio* exposed individually (10 replicates per group) in 4 ml of either water, dimethylsulfoxide control (0.5 $\mu$l of DMSO per ml of water), or 0.5 $\mu$l of a sediment extract (from Pass Fourchon sediment)

a) *S. benedicti* exposed to sediment

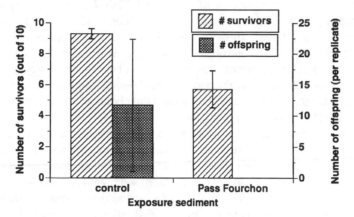

b) *P. pugio* larvae exposed to sediment extract in water

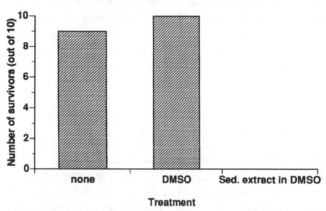

**FIG. 3.** Toxicity of Pass Fourchon sediment: (*a*) Survival (mean number of survivors ± S.E. for three replicates starting with 10 worms each) and reproduction (mean number of offspring ± S.E. per replicate) in *Streblospio benedicti* exposed in the laboratory to control and Pass Fourchon sediment. ANOVA results: $p = 0.041$ for survival (arcsin $\sqrt{p}$ transformed) and $p = 0.274$ for reproduction ($\sqrt{(n + 0.5)}$ transformed). (*b*) Survival of three-day old *Palaemonetes pugio* larvae exposed individually in 10 replicates to estuarine water containing a sediment extract of Pass Fourchon sediment in DMSO (extract of 180 mg sediment in 4 ml total volume), containing DMSO only (2 $\mu$l in 4 ml total volume), or not containing any additions.

in DMSO per ml of water. The grass shrimp exposure was a short-term one (3-day duration), in which no food or aeration were provided. The sediment contamination had toxic effects on all three species. Survival of gobies was significantly reduced in the PF400 sediment (Klerks et al. 1997). In *S. benedicti*, survival was significantly reduced ($p = 0.041$ in ANOVA) in the Pass Fourchon sediment relative to the control group (Fig. 3a). Reproduction was variable (experiment duration was rather short for a reliable determination of reproductive rates) and the difference between the two groups was not statistically significant ($p = 0.274$). However, no reproduction was noted in any of the beakers with the Pass Fourchon sediment. In the experiment with the grass shrimp larvae, all larvae died in the presence of the sediment extract, whereas mortality in the two control groups averaged 5% (Fig. 3b). Later experiments in which adult grass shrimp were exposed to the Pass Fourchon sediment (see below) also demonstrated the toxicity of this sediment to *P. pugio*. These experiments indicate that it is likely that the sediment contamination at Pass Fourchon has exerted a selective pressure on the Pass Fourchon populations.

To determine whether this selective pressure resulted in adaptation to the contamination, laboratory bioassays were conducted for all three species. In these experiments, resistance to PAH-rich Pass Fourchon sediment was quantified in organisms collected in Pass Fourchon (at one or more sites differing in sediment PAH levels) and in those obtained from the nearby control area. Resistance was then compared among these groups using survival analysis (e.g., Collett 1994), allowing the inclusion of right-censored data (results from individuals still alive at the point that the experiment was terminated) without biasing the population resistance estimates.

Two experiments were conducted with the darter goby. Fish were collected from the control area and three Pass Fourchon sites in the first experiment. The number of Pass Fourchon sites was reduced to one in the second experiment, making it possible to substantially increase the sample sizes. Resistance was quantified as survival time in laboratory exposures to Pass Fourchon sediment collected close to the discharge area. Survival analyses did not detect differences in survivorship curves among the groups in either of the two experiments (Fig. 4). There was, therefore, no evidence that *G. boleosoma* had adapted to the contamination at Pass Fourchon.

Experiments with the polychaete *S. benedicti* were conducted with individuals collected from the control site and from three sites along the Pass Fourchon pollution gradient. These worms were exposed to Pass Fourchon sediment in 4 replicates with 9–10 worms each (39 worms per group) in beakers containing 100 ml sediment and 200 ml of overlying water (15°/oo salinity) that was aerated slowly throughout the experiment. Survival was determined after 43 days. Again, no adaptation was observed; rather, the reverse was observed. A regression analysis showed a negative dependence of survival on total PAH levels at the collection site (Fig. 5). The latter pattern is indicative of stress (and thus the presence of a selective pressure for contamination resistance) at the contaminated sites.

The occurrence of adaptation to the contamination at Pass Fourchon was also evaluated for the grass shrimp *P. pugio*. Similar to the situation with the darter goby, two experiments were conducted in which individuals were collected in the control area and in Pass Fourchon. Their resistance to Pass Fourchon sediment was subsequently compared in laboratory bioassays. Grass shrimp were fed flake fish food once per three days during the exposures. In the first experiment, exposures were conducted in 1-l beakers with 100 ml sediment and 700 ml ambient water from the control area. This experiment was terminated on day 40, when control mortality (mortality of individuals from the various sites kept in the lab in beakers with control area

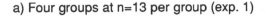

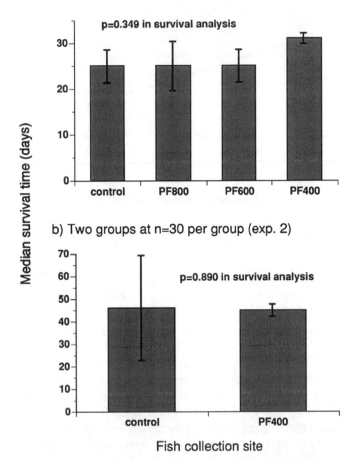

**FIG. 4.** Median survival times (±S.E.) of *Gobionellus boleosoma* collected from the control area and various sites in Pass Fourchon ("PF," with the number following PF indicating the distance in meters from the discharge area). Results of survival analyses comparing survivorship curves among the groups using the Wilcoxon test (e.g., Collett 1994) are shown above graphs. Fish were exposed in the laboratory to PAH-rich Pass Fourchon sediment. (*a*) Experiment one with fish from control area and three Pass Fourchon sites exposed to sediment with total PAH levels of 15,230 μg/g. (*b*) Experiment two with fish from the control area and one Pass Fourchon site. (Reprinted with permission from Klerks et al. 1997).

sediment) averaged 7.5%, whereas mortalities in the experimental groups averaged 78%. Results provided some indication that the individuals collected closest to the discharge area (at PF400) were possibly more resistant to the PAH-rich sediment than the ones collected from the other sites (Fig. 6a). A subsequent experiment was therefore designed to compare resistance between control area individuals and ones from PF400 using larger sample sizes. This experiment was conducted in 400-ml

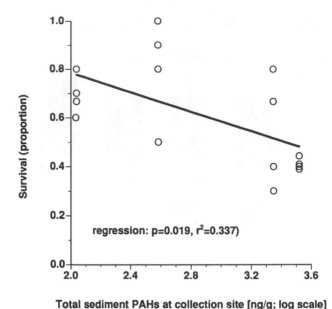

**FIG. 5.** *Streblospio benedicti* survival (proportion surviving a 43-day exposure to Pass Fourchon sediment (<500 μm size and 10,100 ng/g total PAH level) as a function of total PAH levels at the collection site). Exposures were started with four replicates (of nine–ten worms each) per site. Survival proportions on graph are untransformed; PAH concentrations are log-transformed. Regression computed using arcsin $\sqrt{p}$ transformed survival data: $p = 0.040$, $r^2 = 0.268$.

beakers with 50 ml sediment (collected approximately 300 m from the discharge area) and 200 ml ambient water. On day 49, when this experiment was terminated, control mortality was 5%, whereas mortalities in the experimental groups were 45 and 54% for control area and PF400 grass shrimp, respectively. Survival analysis of the results of this experiment did not detect a significant difference in survival time between the two groups (Fig. 6b).

## DISCUSSION

In this study, no evidence of adaptation was found in the four species inhabiting two environments contaminated with a large number of different toxicants. This result contrasts with those obtained in another of this author's studies, in which the occurrence of adaptation was investigated in two species (an oligochaete and a chironomid) inhabiting an area contaminated with mainly one toxicant (cadmium). In the latter study, the oligochaete was found to have evolved resistance to the cadmium while no adaptation was evident in the chironomid species (Klerks and Levinton 1989b; Klerks and Levinton 1993). This dataset, even when combined with the small number of other studies on adaptation to contaminants, is of course too limited to allow any solid conclusions about the importance of contaminant complexity on the potential for

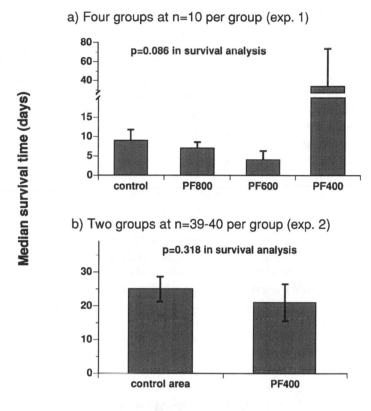

**FIG. 6.** Median survival times (±S.E.) of *P. pugio* collected from the control area and various sites in Pass Fourchon ("PF," with the number following PF indicating the distance in meters from the discharge area). Results of survival analyses comparing survivorship curves among the groups using the Wilcoxon test are shown above graphs. Grass shrimp were exposed in the laboratory to PAH-rich Pass Fourchon sediment during two experiments, with individuals from: (*a*) the control area and three Pass Fourchon sites; (*b*) the control area and one Pass Fourchon site.

the evolution of resistance to contamination, but it forms a starting point in the accumulation of empirical data on this issue. These data are consistent with the theoretical considerations that indicate that the development of resistance will probably become less likely if the number of contaminants increases. It is also consistent with the pattern that such genetic adaptations have generally been reported from populations in environments with a small number of different contaminants (see "Introduction").

Direct comparisons of rates of the development of resistance for individual stressors to those for combinations of stressors have been done for pesticide and antibiotic resistance. A variety of studies reported evidence that the rate of evolution of resistance in insect populations decreases when the number of pesticides is increased

(Pimentel and Bellotti 1976; Macdonald et al. 1983a,b; Georghiou 1990; Immaraju et al. 1990; Huang et al. 1994). Similar results were obtained in two studies on bacterial resistance to antibiotics, showing that combinations of three or four antibiotics were highly effective in preventing the development of resistance to these drugs, whereas resistance to the individual antibiotics evolved rapidly (Carpenter et al. 1945; Klein and Kimmelman 1947). However, other studies on pesticide resistance indicate that results may be more variable (Pimentel and Burgess 1985; Georghiou and Taylor 1986; Tabashnik 1989). In agreement with predictions made in the earlier section on theoretical considerations, there appears a general consensus in the pesticide resistance field that the presence of cross-resistance (or genetic correlations) is the major deciding factor for the effectiveness of using multiple pesticides for retarding the development of resistance. It appears that the strategy is not effective with strong positive correlations in resistance between the different pesticides, but it is very effective when there are strong negative correlations (Georghiou 1990).

More results from selection experiments and field surveys are needed for further confirmation of the theoretical expectations for the effect of contamination complexity on the potential for the development of resistance. However, much of the theoretical and empirical evidence indicates that the evolution of resistance to environmental contamination becomes less likely as more contaminants are involved. Considering that environments that become severely polluted are usually subjected to a variety of contaminants, the continued existence of most of the populations in such environments would thus, seem to be unlikely.

## ACKNOWLEDGMENTS

I thank Joe Neigel and two anonymous reviewers for their insightful comments on earlier versions of this chapter. This research was funded in part by grants from the US Department of Energy (EPSCoR Program) and the Louisiana Education Quality Support Fund.

## REFERENCES

Anderson, G. 1985. Species profiles: Life histories and environmental requirements of coastal fishes and invertebrates (Gulf of Mexico)—grass shrimp. *US Fish and Wildlife Service Biological Services* Rep. 82(11.35). Vicksburg, MS: U.S. Army Corps of Engineers.

APHA—American Public Health Association, American Water Works Association, Water Environment Federation. 1995. *Standard Methods for the Examination of Water and Wastewater*, 19th ed. Washington, DC: American Public Health Association.

Benton, M. J., S. A. Diamond, and S. I. Guttman. 1994. A genetic and morphometric comparison of *Helisoma trivolvis* and *Gambusia holbrooki* from clean and contaminated habitats. *Ecotoxicology and Environmental Safety* 29:20–37.

Bryan, G. W., and L. G. Hummerstone. 1971. Adaptation of the polychaete *Nereis diversicolor* to sediments containing high concentrations of heavy metals. I. General observations and adaptation to copper. *Journal of the Marine Biological Association (UK)* 51:845–863.

Bryan, G. W., and L. G. Hummerstone. 1973. Adaptation of the polychaete *Nereis diversicolor* to estuarine sediments containing high concentrations of zinc and cadmium. *Journal of the Marine Biological Association (UK)* 53:839–857.

Bulmer, M. G. 1989. Maintenance of genetic variability by mutation-selection balance: A child's guide through the jungle. *Genome* 31:761–767.

Carle, K. J., and P. A. Hastings. 1982. Selection of meiofaunal prey by the darter goby, *Gobionellus boleosoma* (Gobiidae). *Estuaries* 5:316–318.

Caro, T. M., and M. K. Laurenson. 1994. Ecological and genetic factors in conservation: A cautionary tale. *Science* 263:485–486.

Carpenter, C. M., J. M. Bahn, H. Ackerman, and H. E. Stokinger. 1945. Adaptability of *Gonococcus* to four bacteriostatic agents, sodium sulfathiazole, rivanol lactate, promin, and penicillin. *Proceedings of the Society for Experimental Biology and Medicine* 60:168–171.

Chen, X. 1993. Comparison of inbreeding and outbreeding in hermaphroditic *Arianta arbustorum* (L.) (land snail). *Heredity* 71:456–461.

Cilek, J. E., and F. W. Knapp. 1993. Enhanced diazinon susceptibility in pyrethroid-resistant horn flies (Diptera: muscidae): Potential for insecticide resistance management. *Journal of Economic Entomology* 86:1303–1307.

Collett, D. 1994. *Modelling survival data in medical research.* London: Chapman and Hall.

DeLaune, R. D., R. P. Gambrell, J. H. Pardue, and W. H. Patrick, Jr. 1990. Fate of petroleum hydrocarbons and toxic organics in Louisiana coastal environments. *Estuaries* 13:72–80.

Diamond, S. A., J. T. Oris, and S. I. Guttman. 1995. Adaptation to fluoranthene exposure in a laboratory population of fathead minnows. *Environmental Toxicology and Chemistry* 14:1393–1400.

Donker, M. H., and C. G. Bogert. 1991. Adaptation to cadmium in three populations of the isopod *Porcellio scaber*. *Comparative Biochemistry and Physiology* 100C:143–146.

Falconer, D. S. 1981. *Introduction to quantitative genetics.* London: Longman Group.

Forbes, V. E. 1996. Chemical stress and genetic variability in invertebrate populations. *Toxicology and Ecotoxicology News* 3:136–141.

Frati, F., P. P. Fanciulli, and L. Posthuma. 1992. Allozyme variation in reference and metal-exposed natural populations of *Orchesella cincta* (Insecta: Collembola). *Biochemical and Systematic Ecology* 20:297–310.

Georghiou, G. P. 1990. Overview of insecticide resistance. In *Managing resistance to agrochemicals*, eds. M. B. Green, H. M. LeBaron, and W. K. Moberg, 18–41. Washington, DC: American Chemical Society.

Georghiou, G. P., and C. E. Taylor. 1986. Factors influencing the evolution of resistance. In *Pesticide resistance: Strategies and tactics for management*, ed. National Research Council Committee on Strategies for the Management of Pesticide Resistant Pest Populations, 157–169. Washington, DC: National Academy Press.

Gromko, M. H., A. Briot, S. C. Jensen, and H. H. Fukui. 1991. Selection on copulation duration in *Drosophila melanogaster*: Predictability of direct response versus unpredictability of correlated response. *Evolution* 45:69–81.

Hill, W. G. 1982. Rates of change in quantitative traits from fixation of new mutations. *Proceedings of the National Academy of Science USA* 79:142–145.

Hoffmann, A. A., and M. W. Blows. 1993. Evolutionary genetics and climate change: Will animals adapt to global warming? In *Biotic interactions and global change*, eds., P. M. Kareiva, J. G. Kingsolver, and R. B. Huey, 165–178. Sunderland, MA: Sinauer Associates.

Hoffmann, A. A., and P. A. Parsons. 1991. *Evolutionary genetics and environmental stress.* Oxford: Oxford University Press.

Holloway, G. J., R. M. Sibly, and S. R. Povey. 1990. Evolution in toxin-stressed environments. *Functional Ecology* 4:289–294.

Houle, D. 1991. Genetic covariance of fitness correlates: What genetic correlations are made of and why it matters. *Evolution* 45:630–648.

Huang, H., Z. Smilowitz, M. C. Saunders, and R. Weisz. 1994. Field evaluation of insecticide application strategies on development of insecticide resistance by Colorado potato beetle (Coleoptera: Chrysomelidae). *Journal of Economic Entomology* 87:847–857.

Immaraju, J. A., J. G. Morse, and R. F. Hobza. 1990. Field evaluation of insecticide rotation and mixtures as strategies for citrus thrips (Thysanoptera: Thripidae) resistance management in California. *Journal of Economic Entomology* 83:306–314.

Jarne, P., and B. Delay. 1990. Inbreeding depression and self-fertilization in *Lymnaea peregra* (Gastropoda: Pulmonata). *Heredity* 64:169–175.

Klein, M., and L. J. Kimmelman. 1947. The correlation between the inhibition of drug resistance and synergism in streptomycin and penicillin. *Journal of Bacteriology* 54:363–370.

Klerks, P. L. 1989. Adaptation to metals in animals. In *Heavy metal tolerance in plants: Evolutionary aspects*, ed. A. J. Shaw, 313–321. Boca Raton: CRC Press.

Klerks, P. L., P. L. Leberg, R. F. Lance, D. J. McMillin, and J. C. Means. 1997. Lack of development of pollutant-resistance or genetic differentiation in darter gobies (*Gobionellus boleosoma*) inhabiting a produced-water discharge site. *Marine Environmental Research* 44:377–395.

Klerks, P. L., and S. A. Lentz. 1998. Resistance to lead and zinc in the western mosquitofish *Gambusia affinis* inhabiting contaminated Bayou Trepagnier. *Ecotoxicology* 7:11–17.

Klerks, P. L., and J. S. Levinton. 1989a. Effects of heavy metals in a polluted aquatic ecosystem. In *Ecotoxicology: Problems and approaches*, eds. S. A. Levin, M. A. Harwell, J. R. Kelly, and K. D. Kimball, 41–67. New York: Springer-Verlag.

Klerks, P. L., and J. S. Levinton. 1989b. Rapid evolution of metal resistance in a benthic oligochaete inhabiting a metal-polluted site. *Biological Bulletin* 176:135–141.

Klerks, P. L., and J. S. Levinton. 1993. Evolution of resistance and changes in community composition in metal-polluted environments: A case study on Foundry Cove. In *Ecotoxicology of metals in invertebrates*, eds. R. Dallinger and P. S. Rainbow, 223–241. Boca Raton, FL: CRC Press.

Klerks, P. L., and J. S. Weis. 1987. Genetic adaptation to heavy metals in aquatic organisms: A review. *Environmental Pollution* 45:173–205.

Kopp, R. L., S. I. Guttman, and T. E. Wissing. 1992. Genetic indicators of environmental stress in central mudminnow (*Umbra Limi*) populations exposed to acid deposition in the Adirondack Mountains. *Environmental Toxicology and Chemistry* 11:665–676.

LADEQ. 1989. *Impact Assessment of Bayou Trepagnier*. Baton Rouge: Louisiana Department of Environmental Quality, Office of Water Resources, Water Pollution Control Division.

Latter, B. D. H., and J. C. Mulley. 1995. Genetic adaptation to captivity and inbreeding depression in small laboratory populations of *Drosophila melanogaster*. *Genetics* 139:255–266.

Leberg, P. L. 1992. Effects of population bottlenecks on genetic diversity as measured by allozyme electrophoresis. *Evolution* 46:477–494.

LeBlanc, G. A. 1982. Laboratory investigation into the development of resistance of *Daphnia magna* Straus to environmental pollutants. *Environmental Pollution (Series A)* 27:309–322.

Lynch, M., W. Gabriel, and A. M. Wood. 1991. Adaptive and demographic responses of plankton populations to environmental change. *Limnology and Oceanography* 36:1301–1312.

Lynch, M., and R. Lande. 1993. Evolution and extinction in response to environmental change. In *Biotic interactions and global change*, eds. P. M. Kareiva, J. G. Kingsolver, and R. B. Huey, 234–250. Sunderland, MA: Sinauer Associates.

Macdonald, R. S., G. A. Surgeoner, K. R. Solomon, and C. R. Harris. 1983a. Development of resistance to permethrin and dichlorvos by the house fly (Diptera: Muscidae) following continuous and alternating insecticide use on four farms. *Canadian Entomologist* 115:1555–1561.

Macdonald, R. S., G. A. Surgeoner, K. R. Solomon, and C. R. Harris. 1983b. Effect of four spray regimes on the development of permethrin and dichlorvos resistance, in the laboratory, by the house fly (Diptera: Muscidae). *Journal of Economic Entomology* 76:417–422.

Murdoch, M. H., and P. D. N. Hebert. 1994. Mitochondrial DNA diversity of brown bullhead from contaminated and relatively pristine sites in the Great Lakes. *Environmental Toxicology and Chemistry* 13:1281–1289.

Nei, M., T. Maruyama, and R. Chakraborty. 1975. The bottleneck effect and genetic variability in populations. *Evolution* 29:1–10.

Nevo, E., T. Shimony, and M. Libni. 1978. Pollution selection of allozyme polymorphisms in barnacles. *Experientia* 34:1562–1564.

Newman, M. C., and M. S. Aplin. 1992. Enhancing toxicity data interpretation and prediction of ecological risk with survival time modeling: An illustration using sodium chloride toxicity to mosquitofish (*Gambusia holbrooki*). *Aquatic Toxicology* 23:85–96.

Pardue, J. H., R. D. DeLaune, and W. H. Patrick, Jr. 1992. Metal to aluminum correlation in Louisiana coastal wetlands: Identification of elevated metal concentrations. *Journal of Environmental Quality* 21:539–545.

Pimentel, D., and A. C. Bellotti. 1976. Parasite-host population systems and genetic stability. *American Naturalist* 110:877–888.

Pimentel, D., and M. Burgess. 1985. Effects of single versus combinations of insecticides on the development of resistance. *Environmental Entomology* 14:582–589.

Posthuma, L., R. F. Hogervorst, and N. M. van Straalen. 1992. Adaptation to soil pollution by cadmium excretion in natural populations of *Orchesella cincta* (L.) (Collembola). *Archives of Environmental Contamination and Toxicology* 22:146–156.

Posthuma, L., and N. M. van Straalen. 1993. Heavy-metal adaptation in terrestrial invertebrates: A review of occurrence, genetics, physiology and ecological consequences. *Comparative Biochemistry and Physiology* 106C:11–38.

Rabalais, N. N., B. A. McKee, D. J. Reed, and J. C. Means. 1991. *Fate and effects of nearshore discharges of OCS produced waters.* Volume II: Technical report. New Orleans, LA: Minerals Management Service, US Department of the Interior, Gulf of Mexico OCS Regional Office.

Rahel, F. 1981. Selection for zinc tolerance in fish: Results from laboratory and wild populations. *Transactions of the American Fisheries Society* 110:19–28.

Ribble, D. O., and J. S. Millar. 1992. Inbreeding effects among inbred and outbred laboratory colonies of *Peromyscus maniculatus. Canadian Journal of Zoology* 70:820–824.

Roff, D. A. 1997. *Evolutionary quantitative genetics.* New York: Chapman and Hall.

Service, P. M., and M. R. Rose. 1985. Genetic covariation among life-history components: The effect of novel environments. *Evolution* 39:943–945.

Sheridan, A. K., and J. S. F. Barker. 1974. Two-trait selection and the genetic correlation. II. Changes in the genetic correlation during two-trait selection. *Australian Journal of Biological Sciences* 27:89–101.

Simberloff, D. 1988. The contribution of population and community biology to conservation science. *Annual Review of Ecology and Systematics* 19:473–511.

Tabashnik, B. E. 1989. Managing resistance with multiple pesticide tactics: Theory, evidence, and recommendations. *Journal of Economics and Entomology* 82:1263–1269.

Toepfer, C. S., and J. W. Fleeger. 1995. Diet of juvenile fishes *Citharichthys spilopterus, Symphurus plagiusa,* and *Gobionellus boleosoma. Bulletin of Marine Science* 56:238–249.

Travis, J., and D. J. Futuyma. 1993. Global change: Lessons from and for evolutionary biology. In *Biotic interactions and global change,* eds. P. A. Kareiva, J. G. Kingsolver, and R. B. Huey, 251–263. Sunderland, MA: Sinauer Associates.

Van Noordwijk, A. J., and W. Scharloo. 1981. Inbreeding in an island population of the great tit. *Evolution* 35:674–688.

Velázquez, A., H. Andreu, N. Xamena, A. Creus, and R. Marcos. 1987. Accumulation of drastic mutants in selection lines for resistance to the insecticides dichlorvos and malathion in *Drosophila melanogaster. Experientia* 43:1122–1123.

Weber, K. E., and L. T. Diggins. 1990. Increased selection response in larger populations. II. Selection for ethanol vapor resistance in *Drosophila melanogaster* at two population sizes. *Genetics* 125:585–597.

Genetics and Ecotoxicology
Edited by V. E. Forbes
Copyright © 1999 Taylor & Francis

# 7

# Natural Selection in Contaminated Environments: A Case Study Using RAPD Genotypes

Christopher W. Theodorakis and Lee R. Shugart

**Abstract.** This work summarizes relevant literature on the occurrence and consequences of pollutant-caused selection and adaptation of tolerance in contaminated populations, including a case study of recent research concerning radionuclide selection on various genotypes of the western mosquitofish (*Gambusia affinis*).

Selection of resistant genotypes in pollutant-exposed populations is relevant to ecological risk assessments for a number of reasons. First, such selection may reduce genetic variation in the population, thus limiting the ability to respond to environmental variation in space and time. Second, development of tolerance may have associated costs in terms of reduction in fitness of individuals and recruitment into populations. Third, migration and gene flow may hinder development of complete resistance in the contaminated populations and increase genetic load in adjacent noncontaminated populations. Finally, the ability for adaptation in opportunistic generalist species and not in specialists may be the mechanism for simplification of communities in contaminated habitats. Hence, increases in resistant genotypes in opportunist populations may signal the potential for changes in community structure.

Differences in genetic composition between populations may suggest selection for specific genotypes, but in order to validate this hypothesis, differential fitness (as reflected by fecundity and/or survival) and possible biochemical or molecular mechanisms for differential responses to toxicants must be shown. In order to do this, populations of mosquitofish (*G. affinis*) were examined using the randomly amplified polymorphic DNA (RAPD) technique. This technique uses the polymerase chain reaction to produce banding patterns similar to other types of DNA fingerprinting. In this work, three previously published studies are summarized and integrated. The first study discovered RAPD markers (visualized as bands in an agarose gel) which were present at an increased frequency in radionuclide-contaminated populations. For the sake of discussion, these markers will be referred to as "contaminant-indicative bands." Furthermore, it was found that fish in the contaminated ponds that displayed the bands had higher fecundity than fish without the bands, suggesting there was a selective advantage associated with these bands. In a second study, fish were collected from a noncontaminated site and caged in a contaminated one. For the most part, fish that displayed the contaminant-indicative bands had a higher survival rate than fish that did not. In the third study, the amount of DNA damage was determined for the different genotypes. The fish were either collected from a population living in the contaminated sites or collected from a noncontaminated site and caged in the contaminated pond or exposed to X-rays in the laboratory. It was found that fish with the contaminant-indicative bands had fewer strand breaks than did fish without such bands. The concordance of all these results indicates that radiation exposure selects for certain genotypes, and the contaminant-indicative bands are markers of genes or other elements that confer a selective advantage in contaminated environments. The significance and applicability of these results to the fields of biomonitoring and ecological risk assessment are discussed.

**Keywords.** Biomonitoring; ecological risk assessments; molecular responses; mosquitofish; radionuclides.

## NATURAL SELECTION IN CONTAMINATED HABITATS: A REVIEW

In order to perform ecological risk assessments of environmental contamination, it is necessary to determine the effects of contaminants on biological systems. Such effects may be measured at the biochemical or molecular level ("suborganismal effects") or at the population, community, or ecosystem levels of organization ("supraorganismal effects"). There are numerous examples of both suborganismal and supraorganismal effects in the literature. Suborganismal effects can be manifested as alterations in macromolecular structure and function. These can include, but are not limited to, induction of detoxification mechanisms such as mixed function oxidase enzymes (Bucheli and Fent 1995) or metallothioneins (Roesijadi 1992), production of DNA lesions (Depledge 1996; Shugart and Theodorakis 1994; Shugart 1993), alterations of chromosomal, cytosolic or stress protein expression (Sanders 1996; Bradley et al. 1994; Theodorakis et al. 1992), inhibition of constitutive enzyme activity (Baturo and Lagadic 1996; Kubitz et al. 1995; Sulatos 1994), and lipid peroxidation (Ahmad 1995). These types of effects are often referred to as biochemical or molecular "biomarkers." Use of these biomarkers is advantageous because they are relatively easy to quantify and can be apparent immediately or soon after the onset of exposure. For example, DNA strand breakage can be detected within one week (Theodorakis et al. 1992) or even within days or hours (Black et al. 1996) of genotoxicant exposure. However, their ecological significance and relevance to environmental risk assessment are often unclear (Shugart chapter 8). For one thing, the physiological consequences of such macromolecular perturbations at even the individual level are not well defined. In most cases, the level of perturbation necessary to produce detrimental effects on organismal health is not fully understood. Finally, in some cases suborganismal effects can be observed when supraorganismal effects are much less apparent. For example, in Oak Ridge, TN, there are several sites that are heavily contaminated with radionuclides, and the fish living in these sites have elevated amounts of DNA strand breakage (Theodorakis et al. 1997), but there seems to be no apparent effect on population density (Trabalka and Allen 1977). For the most part, the ecological consequences of alterations of molecular physiology have yet to be elucidated.

The significance of supraorganismal effects to environmental risk assessment is more apparent, but quantifying such effects is often difficult. Measurement of ecological parameters often entails very time- and labor-intensive efforts (e.g., see Krebs 1989). For instance, determination of population density usually requires multiple sampling efforts, or assumptions (often difficult to verify) about capture success rate or efficiency of sampling methods (Krebs 1989). When measuring such parameters as population dynamics/demographics, community structure, or ecosystem productivity, allowances have to be made for such abiotic variables as habitat structure and complexity, nutrient levels, temperature, etc., which entail further time- and labor-intensive sampling efforts. Additionally, the onset of supraorganismal effects may be latent to the exposure, further complicating determination of causality. Consequently, demonstration of differences in ecological parameters between two or more sites is often problematic, and assignment of a xenobiotic etiology even more so.

Thus it would be advantageous to develop assays which could be measured at the molecular or biochemical level but reflect ecological processes. One such possibility would be examination of population genetic structure, which has been gaining increasing attention in environmental monitoring programs for several reasons. First, exposure to xenobiotic compounds may lead to population crashes, resulting in genetic bottlenecks, i.e., stochastic reductions in genetic diversity—even though the population may subsequently recover. This may limit the population's ability to adapt to changing environmental conditions and could ultimately lead to its extinction (Land and Shannon 1996; cf. Snell et al. chapter 9). It has also been found that populations with less variability grow more slowly than more genetically variable populations (Albert 1993), which could affect their ability to recover from disturbance or other natural perturbations. These changes are of concern because they may persist long after remediation or cessation of the pollutant exposure.

Another way in which exposure to xenobiotics may alter population genetic structure and variability is through selection of contaminant-resistant genotypes (Shaw, chapter 2; Postma and Groenendijk, chapter 5). Adaptation to contaminant stress is apparent through increased resistance of adapted organisms to the effects of contaminant exposure. Such resistance is evident from differential mortality rates or physiological responses between adapted (tolerant) and nonadapted (sensitive) individuals during toxicant exposure (although some may argue otherwise, for the purposes of this paper, the terms "tolerant" and "resistant" will be used as synonyms, as will the terms "sensitive" and "susceptible"). Effects of selection and adaptation merit consideration in ecological risk assessment because 1) selection may affect genetic variability and adaptive potential of the population, 2) adaptation to pollution may complicate extrapolation and comparison of laboratory data to field situations, 3) evidence of pollutant-induced selection could give an indication of perturbations of population recruitment and survivorship, 4) there may be costs associated with adaptation that could compromise the fitness of tolerant individuals, 5) gene flow between contaminated and non-contaminated populations could be deleterious to both, 6) selection for tolerance to one chemical may lead to cotolerance for others, 7) selection could affect patterns of bioaccumulation in tolerant populations, and 8) there could be correlations between selection of resistant genotypes and community or higher level ecological effects.

## Selection and Genetic Variability

Exposure to toxic chemicals may lead to a decrease in genetic variability, not only through reductions of population size, but through selection of tolerant genotypes (or subsequent elimination of susceptible genotypes) (cf. Snell et al., chapter 9). This may lead to a situation in which organisms that are adapted to such exposure may be less well adapted to other stressors. For example, individuals from a population of western mosquitofish (*G. affinis*) that had been living in a contaminated impoundment for over 20 years were shown to have less thermal tolerance than reference populations. This pattern was seen in the $F_2$ generation of lab-reared fish, indicating a genetic basis (Trabalka and Allen 1977). Another investigation found that killifish (*Fundulus hete-roclitus*) that were tolerant to methylmercury were less tolerant to salinity stress than mercury-sensitive individuals (Weis et al. 1982). The ability to respond to variations in temperature and salinity is an important component of the physiological ecology of killifish and mosquitofish. Both are top-feeding fish often found in shallow water, where temperatures can fluctuate dramatically, or in estuarine environments, where salinity can be spatially and temporally variable. Therefore, selection for pollutant-tolerant genotypes may limit the ability of natural populations to respond to novel environmental stressors, both natural and anthropogenic, or to environmental variability.

## Laboratory Versus Field Experiments

It is possible that adaptation may complicate extrapolation and comparison of laboratory tests to field situations because traditional risk assessments use organisms that are collected from noncontaminated populations. It has been found that organisms collected from contaminated sites are affected to a lesser degree than predicted from laboratory exposures (Duncan and Klaverkamp 1983; van Loon and Beamish 1977). Because this pattern may persist in the $F_2$ generation of laboratory-reared organisms (Duncan and Klaverkamp 1983), it can be attributable to genetic adaptation.

Differences in the magnitude of response between laboratory and field populations can also be seen for other types of responses, such as biomarker expression. Theodorakis et al. (1992) found that when the biomarker responses of sunfish (*Lepomis* spp.) exposed to contaminated sediment in the laboratory were compared to feral fish sampled from the location where the sediment was collected, the basic patterns were the same, but the magnitude of response differed. This could be due to differential responsiveness of adapted (feral) and nonadapted (laboratory) populations. For example, killifish (*F. heteroclitus*) were collected from industrially-polluted and reference sites and administered 2378 tetrachlorodibenzo(p)dioxin in the laboratory. It was found that EROD (ethoxyresorufin-o-deethylase) activity, a measure of cytochrome $P_{450}$ activity, differed between the two populations in terms of baseline levels and degree of inducibility (Prince and Cooper 1995). Hence, ecological risk assessments based on extrapolation of both toxicity and biomarker responses from the laboratory to field situations may be confounded by adaptation to contaminant stress (see also Baird and Barata, chapter 11, regarding use of tolerant genotypes in laboratory testing).

## Recruitment and Survivorship

Measurement of recruitment and survivorship in contaminated populations is critical to assessing population-level impacts, but in natural populations these processes are often difficult to quantify. Because selection is often a result of perturbations in recruitment and survivorship (Bickham and Smolen 1994), monitoring the effects of selection (e.g., genotype frequencies) may be one way of assessing potential impacts of pollution on these parameters.

## Costs of Adaptation

There could be costs associated with adaptation to anthropogenic contamination because tolerant genotypes seem to be at a selective disadvantage in noncontaminated environments. This is indicated by the fact that the degree of pollution tolerance per individual or frequency of pollutant-tolerant genotypes rapidly diminishes with distance from the contaminated area (Macnair et al. 1993; Shaw chapter 2). Additionally, when the contamination is removed, populations dominated by tolerant genotypes rapidly revert to being dominated by susceptible genotypes within a few generations (Klerks and Levinton 1989; Luoma 1977, and refs. therein). Evidence of physiologically-based costs of tolerance has also been provided by findings that tolerant individuals reared in clean environments have slower growth (Weis et al. 1982), increased larval development time, reduced hatching success, and increased mortality (Postma et al. 1995; Postma and Groenendijk chapter 5) than more susceptible organisms. There could be several possible explanations for these costs of adaptation, including bioenergetic constraints, pleiotropy, or genetic linkage-associated phenomena. In terms of bioenergetic constraints, production of proteins involved in detoxification or cellular repair could be energetically costly. Therefore, allocation of limited resources to production of these proteins may result in decreased availability of resources for growth and reproduction (Postma et al. 1995). Secondly, pleiotropism could result in an undesired effect on the physiology of constitutive metabolic pathways as a result of increases in expression or activity of xenobiotic-responsive enzymes. Finally, genetic linkage could also contribute to the costs of adaptation if alleles or isoforms of pollutant-responsive genes, which are more efficient at detoxifying contaminants, are linked to alleles or isoforms of "housekeeping genes" (constituitively-expressed genes involved in intermediary metabolism), which are less efficient during routine metabolism, or if selection for tolerant genotypes leads to the breakup of coadapted gene complexes optimized for noncontaminated habitats. As a consequence, individuals that are adapted to polluted environments may still have a lower fitness (as reflected by reproduction and/or growth rate) than organisms in clean environments.

## Gene Flow Between Populations

If the tolerant genotypes are at a disadvantage in clean environments, then migration (gene flow) between contaminated and non-contaminated sites could hinder adaptation to pollution in the contaminant-exposed populations (e.g., Postma and Groenendijk, chapter 5). This would be particularly true if the progeny of crosses between tolerant and sensitive individuals were more susceptible to the toxicant than either parental line. For example, it has been found in the mussel, *Mytilus edulis*, that such progeny had more developmental abnormalities when exposed to copper than did embryos of pure tolerant or pure sensitive crosses (Hoare et al. 1995). Alternatively, gene flow from contaminated sites into noncontaminated areas may increase the genetic load of nearby noncontaminated populations because 1) there would be a continuous influx of suboptimal genotypes or 2) there still may be deleterious mutations present in the pollution-tolerant population (Depledge 1996). Thus, the ecological effects of contamination may extend beyond the contaminated zone, and migration between contaminated and noncontaminated sites may compromise the fitness of individuals from both.

An alternative scenario would be if individuals immigrated from a chronically polluted site into a newly polluted site. Such an instance could occur, for example, via the spread of contamination or increased pollutant load into the ecosystem. In this case gene flow would enhance, rather than retard, the rate and degree of divergence of the newly contaminated population from its ancestral gene pool. This could also mitigate effects of pollution on the population because the loss of sensitive individuals in the newly polluted sites could be offset by influx of pre-adapted individuals.

## Adaptation and Cotolerance

Occasionally, organisms that are adapted to one chemical will exhibit increased tolerance to another, a phenomenon often referred to as "cotolerance." This phenomenon has long been recognized in plants exposed to heavy metals. Individuals from populations exposed to a particular metal for several generations (either in the laboratory or in the field) will sometimes develop resistance to other metals as well (Whitton and Shehata 1982; Cox and Hutchinson 1979; Allen and Shepard 1971). However, this is not true in all cases (Whitton and Shehata 1982), and, in fact, selection for increased resistance to one metal sometimes resulted in an increased sensitivity to others (Whitton and Shehata 1982). Cotolerance has also been found for other contaminants besides metals. Algae communities adapted to arsenate stress were also resistant to other photosynthetic inhibitors (Blank and Wängberg 1991), and grasses that were selected for s-triazine resistance were tolerant of other herbicides as well (Erickson et al. 1985). Another study found that fathead minnows (*Pimephales promelas*) selected for fluoranthene resistance were also more resistant to the effects of acute copper exposure than were nonselected individuals (Schuleter et al. 1997). Thus, it has been demonstrated in both plants and fish that selection for resistance to certain chemicals may lead to resistance to others.

The mechanisms by which cotolerance develops have not been determined to any great extent, but cotolerance typically occurs for chemicals which are structurally similar (Blank and Wängberg 1991) and/or have similar or overlapping modes of action (Blank and Wängberg 1991; Erickson et al. 1985). Also, xenobiotic tolerance may involve detoxification mechanisms (e.g., metallothionein, cytochrome $P_{450}$) or changes in life history characteristics (Mulvey et al. 1995, Postma et al. 1995) that are not specific to any one chemical. Thus, adaptations that confer resistance to one chemical may prove advantageous when the organism is exposed to other chemicals.

The occurrence of cotolerance suggests that selective pressures imposed by one chemical can predispose an organism to be tolerant of another (cf. Klerks, chapter 6). Consequently, when populations are chronically exposed to contaminants, the evolution and ultimate fate of the population (e.g. extinction versus persistence) may be influenced by previous exposure to chemicals with similar modes of action. These factors need to be taken into account when assessing the possible effects of pollution on population genetics and their implications for ecological risk.

## Adaptation and Bioaccumulation

Contaminant exposure often selects for individuals that are able to tolerate elevated levels of contaminants through enhanced sequestration of toxic compounds (Postma et al. 1995; Klerks and Bartholomew 1991). This could lead to a situation in which the tolerant organisms have greater body burdens than sensitive ones. Alternatively, genetic adaptation to contaminant stress may involve selection for individuals which are more effective at excluding or eliminating toxicants from the body (Posthuma et al. 1992), which could result in tolerant organisms having a proportionally lower body burden than sensitive ones. In this case, risk assessments based on body burdens of tolerant species may underestimate levels of contaminant exposure to or body burdens in other species. As of yet, though, the relationships between genetic adaptation and bioaccumulation and their implications for ecological risk assessments have not been investigated. Consequently, any research in this area is urgently needed.

## Community Level Effects

Communities residing in polluted habitats include tolerant populations and often have fewer species (Luoma 1977). This is most likely due to the fact that not all species are equally adept at responding to the effects of contaminant selection. Many species that respond to such selection are generalized opportunists, which typically are more genetically plastic than more specialized organisms. Simplification of community structure may then occur as specialized species are replaced by more adaptable generalists (Luoma 1977). Also, some species may not have the ability to adapt to such selective pressures (Macnair et al. 1993) because this depends upon the presence of tolerant genotypes in pre-exposed populations. Other species may be unable to respond to selection due to small population sizes. This situation exists in many wildlife (Hubert and Luiker 1996) and endangered species. Selection is also more likely to occur in small, short-lived species with high reproductive output and short generation times. Larger, longer-lived organisms may be less likely to adapt and thus more at risk of extinction. An additional consideration is that many pollutants tend to biomagnify up the food chain, so that larger predators may receive higher doses and have greater body burdens, which could lead to stronger selection pressures than those experienced by smaller organisms at lower trophic levels. Consequently, even though adaptation occurs in some species, this does not mean that it would result in the negation of ecological risk in all species.

Additionally, the correlation between adaptation/tolerance in opportunistic species with the simplification of contaminated communities implies that evidence for genetic adaptation in small generalist species could be a tool for assessing the potential for changes in community structure. For instance, evidence of an increase in pollutant-resistant genotypes in small fish populations may serve as an early warning indicator for higher-level effects.

## Demonstration of Contaminant Selection

Demonstration of the occurrence of contaminant selection may be accomplished in a variety of ways. Differences in allele frequencies between contaminated and reference sites have been found to occur in many species (Gillespie and Guttman 1988; chapter 4; Guttman 1994; cf. Weis et al. chapter 3), and this may suggest that there is a selective advantage to certain genotypes over others in contaminated populations. But there are many possible contributing factors to differences in allele frequencies (e.g., neutral mutation and genetic drift) founder effects, gene flow, so this in and of itself is not conclusive evidence that contaminant-induced selection influences population genetic structure. One way to provide validity for this hypothesis would be to perform laboratory experiments in which organisms were exposed to contaminants and then determine if some component of fitness (e.g., fecundity or survival) was dependent on genotype. For instance, it has been found that survival time of eastern mosquitofish (*Gambusia holbrooki*) exposed to heavy metals was correlated with allozyme genotype (Newman et al. 1989). However, this pattern was not consistent for fish sampled from different populations or during different years (Lee et al. 1992; Diamond et al. 1991). In a related study, Gillespie and Guttman (1988) found that there were certain alleles that were more prevalent in contaminated populations of central stoneroller (*Campostoma anomalum*), and that these genotypes had higher survival than others when exposed to copper in the laboratory (Changon and Guttman 1989a). They further went on to show that the enzymatic activity of these particular alleles was less inhibited by copper in vitro than was that of the alleles that were less prevalent in noncontaminated populations (Changon and Guttman 1989b). This study was significant because it not only linked genotype frequencies with a component of selection (survival), but it also demonstrated a biochemical basis for this phenomenon. Because fecundity is another component of selection, studies that demonstrate differential fecundity of genotypes would be of particular value. For example, Mulvey et al. (1995) exposed *G. holbrooki* females to mercury and found differential survival of glucose-phosphate isomerase genotypes. These differences in survival were found to be consistent with differences in reproductive performance in terms of number of gravid females and number of developing embryos per female. Thus, in order to provide definitive evidence for the occurrence of contaminant-induced selection on natural populations, genotype frequencies in the field must be compared with fitness components (e.g., reproduction or survival). Additional evidence could be provided by elucidation of a possible biochemical or molecular basis of the genotype-dependent correlates of fitness.

Therefore, a 1992 study was initiated in an attempt to uncover evidence for xenobiotic-induced selection in *G. affinis* populations exposed to radionuclide contamination using this approach. The genetic assay used in this case was the randomly amplified polymorphic DNA (RAPD) technique.

## CASE STUDY: GENETIC ECOTOXICOLOGY
## OF RADIONUCLIDES IN MOSQUITOFISH

### Introduction

Although the above examples provide convincing evidence for contaminant-induced selection, they do have some limitations for complete understanding of the genetic basis for adaptation to contaminant exposure. For one thing, most of the aforementioned studies rely on allozyme analysis, which only uncovers polymorphisms resulting from detectable changes in isoelectric point of proteins. Furthermore, they represent a small proportion of transcribed genes. There could be potentially hundreds (if not thousands) of other genes involved in pollution resistance that are not detected by this technique.

An alternative approach is to use genes that are responsive to xenobiotic exposure as probes for examination of population genetic structure. One strategy could be to choose a gene that has already been identified as being contaminant-responsive. For instance, Wirgin et al. (1991) found unique polymorphisms in polluted populations of Tomcod (*Microgadus tomcod*) using the cytochrome $P_{450}1A1$ gene. Because of the large number of genes that may be responsive to each class of contaminant, choosing the most appropriate gene could be cumbersome. Another strategy is to examine all transcribed genes and identify ones that are differentially expressed in tolerant and sensitive individuals via cDNA screening of differentially expressed mRNAs (Urwin et al. 1996). These genes could then be used as probes in population genetic studies. Both of these approaches are basically hit-or-miss strategies. Whether one chooses a previously-identified gene or identifies one through RNA screening, it is not known a priori if these probes will reveal polymorphisms that are useful in monitoring environmental contamination. If not, this could result in unfruitful efforts and wasted time.

A third strategy would involve the use of nonlabor intensive techniques for examination of a large number of anonymous loci, identify those polymorphic markers that are able to discriminate between contaminated and reference populations, and then use molecular techniques to isolate and characterize these markers. One such technique that could examine large numbers of loci simultaneously is the randomly amplified polymorphic DNA (RAPD) assay. This technique uses the polymerase chain reaction with short (10 base pair) oligonucleotide primers to produce DNA fragments, which when analyzed by agarose gel electrophoresis, produce banding patterns similar to DNA fingerprints or mitochondrial restriction fragment length polymorphisms (RFLPs). These bands have been found to exhibit Mendelian inheritance typically acting as dominant markers (Welsh and McClelland 1990; Williams et al. 1990). The RAPD technique has been used in recent years to study population subdivision (Kuusipalo 1994), genetic diversity and genetic distance (Naish et al. 1995), species identification (Dinesh et al. 1993b), and construction of genetic linkage maps (Postlethwait et al. 1994) in various fish species. In cases where RAPDs were used with other methods of DNA fingerprinting, similar results were obtained (Naish et al. 1995). However, its widespread use in studies of contaminated populations has yet to be realized. This is unfortunate because the technique has several advantages over other currently used techniques. First, it requires much less DNA and does not use radioactive probes. Second, it encompasses genetic differences that occur at multiple loci. Furthermore, the configuration of RAPD priming sites is similar to genetic elements that may be involved in resistance to ionizing radiation (Schwartz and Vaughn 1989; Dorling and Starlinger 1986), so RAPDs may be appropriate for genotoxic study. Finally, RAPDs can be used to produce genetic markers of certain phenotypic characteristics (Michelmore et al. 1991), and so may provide markers of loci that are responsive to contaminant selection. Thus, RAPDs might be useful in monitoring for the effects of selection and in the identification of markers of contaminant resistance.

There are, however, some drawbacks to the RAPD technique. One is that the genomic identity and possible biological significance of RAPD bands are currently unknown. These bands can be amplified from various locations in the genome, from coding or noncoding regions, or from repetitive or single-copy DNA. Therefore, assigning adaptive significance to particular bands should be viewed with caution, at least until rigorous investigation of the genomic locations and possible functions of the amplification sites have been thoroughly characterized. Secondly, RAPDs are dominant markers, so estimation of population genetic parameters (e.g., heterozygosity, F statistics) may be more biased than with codominant markers such as allozymes or microsatellites (Lynch and Milligan 1994). Furthermore, it has been found that in some cases bands are produced which are not reproducible or are difficult to score. Finally, the RAPD bands are usually identified by molecular weight and not by nucleotide sequence. Therefore, two DNA fragments with similar molecular weights but different sequences may be identified as a single band. These drawbacks are further discussed and reviewed in Hadrys et al. (1992) and Williams et al. (1990). However, if the reactions are conducted in duplicate, if care is taken to avoid cross-contamination of DNA samples, and if the reaction conditions are rigidly standardized, the majority of the banding patterns are consistent and reproducible, and those which are not can easily be eliminated from the analysis. Other studies have also found that RAPD patterns are highly reproducible if proper care is taken (Dinesh et al. 1993a).

Recently, a series of studies (Theodorakis et al. 1997; Theodorakis and Shugart 1997) were conducted using the western mosquitofish (*G. affinis*). These studies used the RAPD technique to investigate the molecular mechanisms of selection and adaptation to genotoxicant, in this case radionuclide, stress. RAPD bands were used to identify punitive markers of loci that are involved in adaptation to genotoxicant stress.

Evidence for such adaptation was inferred from 1) comparison of band frequencies in populations from contaminated and reference populations and 2) fecundity, relative DNA damage, and survival of fish with and without certain RAPD markers when exposed to radiation. The following discourse summarizes and integrates these endeavors in relation to each other and the significance of these findings to biomonitoring and ecological risk assessment.

## DNA Strand Breakage and Reproduction

The first step in examining the possibility of radiation-induced selection was to determine whether there was evidence of a measurable biological response of the fish to radiation exposure and whether this response was related to differential reproductive performance (assumed to be a reflection of fitness). One well-known biological response to radiation exposure is DNA strand breakage (Kirsch et al. 1991; Shulte-Frohlinde and von Sonntag 1985; Roots and Okada 1975; Shugart chapter 8). It has also been shown that when organisms or cultured cells are exposed to ionizing radiation in the laboratory, the organisms or cell lines that are more radioresistant have fewer strand breaks than those which are radiosensitive (Olive 1992). Therefore, if fish in wild populations have differing amounts of radio-resistance, this should be reflected in differential amounts of DNA strand breakage. Furthermore, it has been found that radiation exposure can have detrimental effects on fecundity of fish (Woodhead 1977). If individuals vary in their degree of radio-resistance, then the effect of radiation exposure on fecundity should also vary between organisms. Thus, if there is differential radio-resistance in natural populations of fish, and if relative radio-resistance imparts a selective advantage, then this should be reflected by a relationship between DNA strand breakage and fecundity.

In order to test this hypothesis, DNA strand breakage was first measured in fish exposed to ionizing radiation in situ (Theodorakis et al. 1997). This was done by examination of four populations of mosquitofish: two that were contaminated with radionuclides (Pond 3513 and White Oak Lake) and two that were from clean sites (Crystal Springs and Wolf Creek; Theodorakis et al. 1997). Pond 3513 is a settling basin at the Oak Ridge National Laboratory compound and is heavily contaminated with nuclear fission products (including $^{137}$Cs, $^{90}$Sr, and $^{60}$Co) and small amounts of plutonium and uranium (Tamura et al. 1977). White Oak Lake is an impoundment located on the Department of Energy's Oak Ridge Reservation and is also contaminated with nuclear fission products, as well as with heavy metals, PCBs and hydrocarbons (Blaylock et al. 1991). Crystal Springs is a first order stream that originates from an underground limestone cave in Oak Ridge, TN. Wolf Creek is a first order stream that empties into the Clinch River near Clinton, TN. In 1977, Pond 3513 was colonized with about 250 fish collected from Crystal Springs. Before this, no *Gambusia* had inhabited this pond. All other study sites were naturally inhabited by this species.

DNA was extracted from the livers, and the amount of DNA strand breakage was determined electrophoretically, as described in Theodorakis et al. (1997; see also Shugart chapter 8). The median molecular length (MML) of the DNA after electrophoresis was calculated using DNA molecular length standards. The value of the MML is inversely related to the amount of DNA strand breakage in the sample: if a DNA sample has more strand breaks, the size distribution of the DNA will be skewed more toward smaller DNA fragments and the MML will be lower. Because DNA is a double-stranded molecule, strand breakage can occur on both adjacent strands (double strand breaks) or on one strand and not the other (single strand breaks). If electrophoresis is conducted under alkaline conditions, the DNA is separated into its complimentary strands. In this case the MML will be referred to as the single-strand MML (ssMML), although both single and double strand breaks could result in a lower ssMML. If the electrophoresis is conducted under neutral pH, the DNA will remain in its double-stranded form (double-stranded MML), and the MML will only be affected by double-strand breaks.

The MML of each fish was also compared to its fecundity. Because fecundity in *Gambusia* shows a linear correlation with body size (Blaylock 1969), it is reported as brood size (number of eggs or developing embryos)/body length. It was then determined if there was a correlation between MML and fecundity.

The results of this study demonstrated that the double-strand MML of fish from White Oak Lake and Pond 3513 populations was lower than that from either of the two reference sites (Fig. 1), indicating a higher degree of strand breakage. Also, the single-strand MML in Pond 3513 was lower than in any other population. Note that the integrity for the Crystal Springs site was lower than that for the Wolf Creek site, suggesting that the background concentration of DNA-damaging agents is higher in Crystal Springs. This will be further discussed in the next section.

It was also found that there was a direct correlation of DNA integrity (i.e., MML) with fecundity (Fig. 2) at least for single-strand MML. There were no such relationships in the reference sites. This implies that resistance to DNA damage carries a fitness component, in that individuals that are better able to prevent and/or repair DNA damage are at a selective advantage in the environment. However, it could also be that this relationship is due to environmental factors. If certain fish receive a higher dose of radiation than others, then they would exhibit both an increased level of strand breakage and lower fecundity. Also, fish in poorer physical condition may not be able to repair DNA damage as efficiently as a healthier fish and would also have lower fecundity. In this case the correlation between DNA integrity and fecundity would stem from the fact that they are both dependant on physical condition. Therefore, the population genetics of these fish were examined to determine if this correlation had a genetic, rather than environmental, etiology.

## Population Frequencies of RAPD Markers

For these experiments, the RAPD technique was employed in order to determine if the certain genotypes could impart a selective advantage in contaminated environments. However, the RAPD technique can potentially produce markers of hundreds if not thousands of loci, some that are responsive to contaminant selection and some that

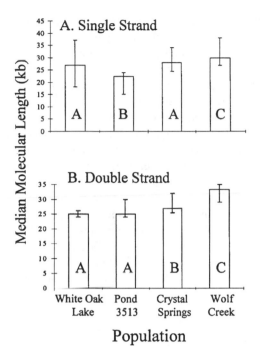

**FIG. 1.** DNA integrity (in terms of median molecular length, MML) for two populations of mosquitofish contaminated with radionuclides (White Oak Lake and Pond 3513) and two reference populations (Crystal Springs and Wolf Creek). Bars represent medians and error bars represent first and third quartiles. Bars labeled with different letters are significantly different ($p < 0.05$, Wilcoxon Rank Sum Test). DNA was examined by electrophoresis run under alkaline in order to separate the DNA into single-strands (a) or neutral conditions to examine DNA in its double-stranded form (b). A lower MML indicates a lower DNA integrity and more DNA strand breaks. (Figure from Theodorakis et al. 1997, with permission.)

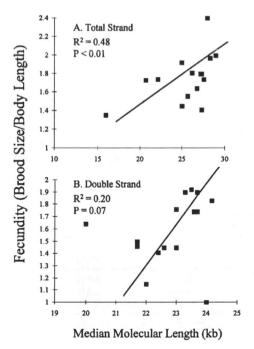

**FIG. 2.** Relationship between DNA integrity (as indicated by median molecular length, MML) and brood size in mosquitofish collected from a radioactively-contaminated pond (Pond 3513). DNA integrity is inversely proportional to the number of DNA strand breaks. (Data taken from Theodorakis et al. 1997, with permission.)

are not. One way to tentatively identify the loci that are responsive to contaminant selection would be to identify bands that are present at a higher frequency in contaminated relative to reference sites. This was accomplished using the four mosquitofish populations described in the previous section. Because the Pond 3513 population was colonized with individuals from Crystal Springs, this provides a unique opportunity to observe any differences in population genetic structure between Pond 3513 and Crystal Springs that are due to possible radionuclide-induced selection. The band frequencies in Pond 3513 and Crystal Springs were compared to those in White Oak Lake and Wolf Creek in an attempt to distinguish between radionuclide selection and other evolutionary processes (e.g., founder effect, genetic drift, etc.) or habitat differences between sites.

Thirty-five to 45 adult female fish were captured from each of these sites. DNA was isolated from the blood and the RAPD technique was performed as described in Theodorakis and Shugart (1997). The PCR products were electrophoresed on 1.5% agarose, stained with ethidium bromide, and photographed under UV light. RAPD bands were scored by eye. Frequencies of each band were calculated as (number of fish with band)/(total number of fish in the sample). Differences in band frequencies were tested with a $\chi^2$ test.

A total of 142 RAPD bands were identified. The frequencies of these bands were then used to calculate genetic distances between populations. It was found that the two most similar populations were Pond 3513 and White Oak Lake, i.e., the smallest genetic distance existed between these populations (Table 1). If the genetic distances were due strictly to nonselective processes, it would be expected that the two most similar populations would be Pond 3513 and Crystal Springs because Crystal Springs was the origin of the Pond 3513 fish. The fact that Pond 3513 and White Oak Lake were the most similar suggests that selection influenced the distribution of RAPD bands within the population.

Of these 142 bands, 17 were found to be present at a higher frequency in the contaminated sites relative to the reference sites. Also, there were four bands that were found to be present at a lower frequency in the contaminated sites relative to the reference sites (Table 2). Subsequent analysis involving fecundity, survivorship, and DNA strand breakage will focus on these particular bands, which, for the sake of discussion, will be referred to as "contaminant-indicative bands." The nomenclature of these bands is derived from the name of the primer used to amplify them (designated

TABLE 1. Pairwise genetic distances[a] between four populations
of mosquitofish

|  | Pond 3513 | Crystal Springs | Wolf Creek |
|---|---|---|---|
| White Oak Lake | 0.189 | 0.276 | 0.307 |
| Pond 3513 | — | 0.282 | 0.319 |
| Crystal Springs | — | — | 0.313 |

[a]A smaller genetic distance indicates that populations are more genetically similar.

**TABLE 2.** *Frequency of selected RAPD bands [a] for two radionuclide-contaminated (Pond 3513 and White Oak Lake) and two noncontaminated (Crystal Springs and Wolf Creek) mosquitofish populations [b]*

| | Population | | | |
|---|---|---|---|---|
| Band | Pond 3513 | White Oak Lake | Crystal Springs | Wolf Creek |
| A. Frequency in Contaminated > Reference | | | | |
| $OPD2_{1590}$ | 1.00 | 0.97 | 0.06 | 0.57 |
| $OPD2_{1060}$ | 1.00 | 1.00 | 0.09 | 0.00 |
| $OPD7_{1390}$ | 0.31 | 0.18 | 0.00 | 0.02 |
| $OPD7_{1160}$ | 0.62 | 0.36 | 0.21 | 0.20 |
| $OPD13_{430}$ | 0.96 | 0.74 | 0.59 | 0.00 |
| $OPD20_{605}$ | 0.20 | 0.67 | 0.06 | 0.11 |
| $OPD20_{430}$ | 0.78 | 0.48 | 0.09 | 0.00 |
| $UBC4_{2060}$ | 0.91 | 0.74 | 0.52 | 0.10 |
| $UBC4_{1610}$ | 0.71 | 0.77 | 0.31 | 0.27 |
| $UBC4_{1410}$ | 0.47 | 0.53 | 0.34 | 0.13 |
| $UBC6_{580}$ | 0.44 | 0.64 | 0.21 | 0.31 |
| $UBC6_{460}$ | 0.20 | 0.23 | 0.00 | 0.00 |
| $UBC9_{920}$ | 0.77 | 0.52 | 0.45 | 0.03 |
| $UBC12_{1270}$ | 0.57 | 0.48 | 0.00 | 0.00 |
| $UBC12_{1160}$ | 0.72 | 0.19 | 0.00 | 0.00 |
| $UBC12_{860}$ | 0.13 | 0.29 | 0.00 | 0.02 |
| $UBC16_{1016}$ | 0.76 | 0.78 | 0.36 | 0.43 |
| B. Frequency in Reference > Contaminated | | | | |
| $OPD8_{560}$ | 0.85 | 0.67 | 0.93 | 0.97 |
| $OPD13_{790}$ | 0.26 | 0.21 | 1.00 | 0.97 |
| $OPD13_{680}$ | 0.48 | 0.53 | 0.83 | 0.97 |
| $OPD13_{260}$ | 0.02 | 0.00 | 0.83 | 0.97 |

[a]The names of the bands are indicated by the names of the primer that was used to amplify them (designated by the manufacturer) followed by the molecular length of the band in subscript. For example, band $OPD7_{1390}$ indicates a band amplified using primer OPD7 that is 1390 base pairs long. Only bands that show consistent differences in terms of contaminated versus reference are reported here.

[b]The sample sizes for each population are as follows: Pond 3513, 40; White Oak Lake, 40; Crystal Springs, 38; Wolf Creek, 35.

by the manufacturer of the primer) followed by the molecular length of the band in subscript. For example, band $OPD7_{1390}$ is a DNA fragment 1390 base pairs (bp) long and amplified from the genomic DNA using primer OPD7.

## Differential Fecundity of RAPD Genotypes

The differences in frequency of the contaminant-indicative bands between contaminated and reference populations suggest they may be markers of loci that provide some sort of selective advantage in radionuclide-contaminated habitats. If this is true, then this should be reflected in some component of fitness. To test this hypothesis, fecundity was examined in fish from each of the four populations with and without the contaminant-indicative bands.

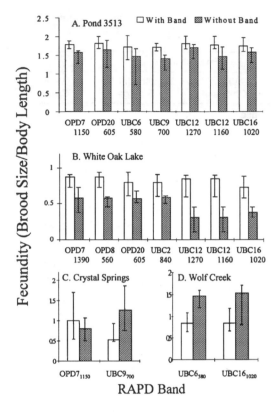

**FIG. 3.** Fecundity of fish with and without selected RAPD bands. Data are presented for fish collected from radioactively-contaminated Pond 3513 (*a*) and White Oak Lake (*b*) and reference sites, Crystal Springs (*c*) and Wolf Creek (*d*). Names of RAPD bands are as explained in Table 2. Data are presented for bands for which the frequency in contaminated populations is greater than that in reference populations (see Table 2) and are presented here only if there is a significant difference between fish with and without the band ($p < 0.05$, Wilcoxon Rank Sum test). Bars and error bars represent medians with first and third quartiles. Asymmetric error bars represent a skewed distribution. Data taken from Theodorakis and Shugart 1997.

It was found that for seven of the contaminant-indicative bands in Pond 3513 and White Oak Lake, females that displayed these bands had a higher fecundity than those without the bands. This was true for only one band in the Crystal Springs population. For band UBC9$_{700}$, the fecundity of the Crystal Springs fish with the bands was less than that of fish without the bands, opposite to the trends in the contaminated populations. For the Wolf Creek population, there were two bands that showed this trend, again, opposite to what was found in the contaminated sites (Theodorakis and Shugart 1997; see Fig. 3).

## Differential Survival of RAPD Genotypes

Another component of fitness is survival. Thus, if there is differential fitness between genotypes, then survival should be dependent on genotype when fish are exposed to radiation. In order to test this hypothesis, fish were collected from a noncontaminated pond and caged in another noncontaminated pond or in Pond 3513. The cages were constructed of metal wire and plastic netting, and were approximately 0.5 m in diameter by 1 m in height. Forty fish were placed in two cages in each of the ponds for two weeks. In the noncontaminated pond, all fish survived, but in Pond

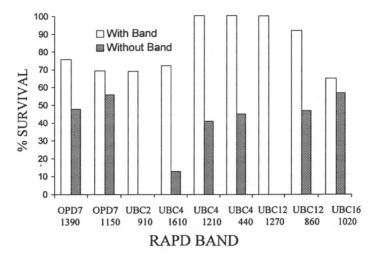

**FIG. 4.** Percent survival of caged fish with and without selected RAPD bands. Fish were collected from a noncontaminated pond and caged in a radioactively-contaminated pond for two weeks. Data taken from Theodorakis and Shugart, 1998b.

3513, only 24 survived. Small samples of caudal fin were collected from each fish in the original sample and from the survivors. In this way, the number of fish with each contaminant-indicative band before and after exposure was determined. Percent survival of fish with each band was calculated as (number of fish with band after caging/number of fish with band before caging) × 100. The percentage of survival of fish without each band was calculated in the same way.

It was found that for nine of the contaminant-indicative bands, the percentage of survival of fish with the band was greater than that for fish without the band (Theodorakis and Shugart 1998b; see Fig. 4). For bands $UBC4_{1210}$ and $UBC4_{440}$, there was 100% survival for fish that displayed this band, and for band $UBC2_{910}$, none of the fish without the band survived. For band $UBC12_{1270}$, not only did all of the fish with the band survive, but none of the fish without the band survived (Fig. 4).

### Differential DNA Damage Between Genotypes

The data above imply that the contaminant-indicative bands may be markers of loci that confer some sort of selective advantage in contaminated populations—in this case, a higher degree of relative radio-resistance. It was mentioned above that the amount of DNA damage could be a reflection of relative radio-resistance. It then follows that the relative amount of DNA damage should be dependent on RAPD genotype. Therefore, the MML was compared for individuals with and without the contaminant-indicative bands.

In order to do this, three separate experiments were performed. The first experiment used fish collected from the four populations described previously and used in determination of band frequencies (Theodorakis and Shugart 1997). In the second experiment, 30 fish were collected from a noncontaminated pond and exposed to 20 Gy (approximately 12 minutes exposure) of X-rays in the laboratory (Theodorakis and Shugart 1998a). The third experiment used the fish from the caging experiment described above (Theodorakis and Shugart 1998b).

In the field experiment, recall that there were 17 contaminant-indicative bands originally identified. Five fish in the Pond 3513 population and four in the White Oak Lake population with these bands had a higher DNA integrity (fewer strand breaks) than did fish without the bands (Fig. 5). This pattern was not reflected in the noncontaminated sites. For the X-ray and caging experiments, only 10 and 11, respectively, of the original 17 bands were examined. The other bands were not present or not polymorphic in the fish used for these experiments. The results from the caging experiment indicated that for five of the bands, the fish that displayed the bands had higher DNA integrity than fish without the bands. For the X-ray experiment, this pattern was also seen for five of these bands (Fig. 5).

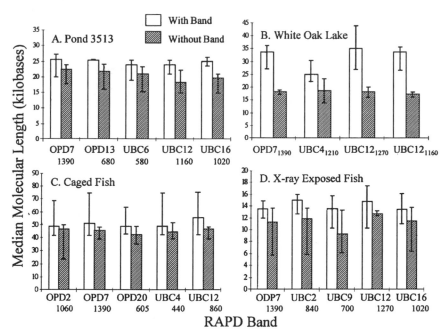

**FIG. 5.** DNA integrity, in terms of median molecular length, for fish with and without selected RAPD bands. Data are for fish collected from radioactively-contaminated pond (*a,b*) or collected from a non-radioactive pond and either caged in a contaminated pond (*c*) or exposed to X-rays in the laboratory (*d*). Data are presented for bands for which the frequency in contaminated populations is greater than that in reference populations (see Table 2) and is presented here only if there is a significant difference between fish with and without the band ($p < 0.05$, Wilcoxon Rank Sum test). Bars and error bars represent medians with first and third quartiles. Asymmetric error bars represent a skewed distribution. Data are taken from Theodorakis and Shugart 1998a (*a,b,d*) or 1998b (*c*).

## Discussion

The results from these experiments indicate that the contaminant-indicative bands are markers of loci that confer some sort of selective advantage in contaminated populations. There is always the possibility the differences in band frequency between contaminated and reference populations may be due to genetic drift, founder effects, neutral mutation, or selection due to a variable other than radiation. However, the similarity of the trends displayed by band frequencies, fecundity, DNA damage, and survival would argue against this possibility. For three bands there is complete concordance between all four of these parameters (Table 3). Unfortunately, the identity and possible function of the loci from which these bands are amplified would have to be elucidated to provide conclusive evidence. Although it would be of interest to determine the molecular identity of all of the contaminant-indicative bands, it would be of special interest to determine the identity of the bands listed in Table 3, and such efforts initially shall focus on these bands.

The presence of an increased frequency of resistant genotypes may also be used to identify sites where contamination is present but was not previously suspected. For example, notice that the population that is genetically most similar (i.e., the

**TABLE 3.** *Summary data for RAPD analysis of mosquitofish living in contaminated or reference ponds**

| RAPD band[e] | Population frequency[a] | | | | Fecundity[b] | | DNA integrity (MML)[c] | | Survival of caged fish[d] | |
|---|---|---|---|---|---|---|---|---|---|---|
| | White Oak Lake | Pond 3513 | Crystal Springs | Wolf Creek | With band | Without band | With band | Without band | With band | Without band |
| UBC4 1610 | 0.77 | 0.71 | 0.31 | 0.27 | 1.81 | 1.64 | 25.4 | 21.6 | 71.8 | 12.5 |
| UBC12 1270 | 0.48 | 0.57 | 0 | 0 | 1.81 | 1.66 | 23.9 | 20.8 | 100 | 0 |
| UBC16 1020 | 0.78 | 0.76 | 0.36 | 0.43 | 1.75 | 1.58 | 23.8 | 20.8 | 64.7 | 56.5 |

*Data are presented for 3 RAPD bands that were present at a higher frequency in contaminated populations than in reference populations.

[a]Frequency of bands in each of 4 different populations. White Oak Lake and Pond 3513 are contaminated with radionuclides, Crystal Springs and Wolf Creek are reference sites.

[b]Fecundity (number of embryos/body length of female) for fish with or without certain RAPD bands. Data are from fish living in Pond 3513.

[c]DNA integrity for fish with and without certain RAPD bands. Data are from fish living in Pond 3513. DNA integrity is measured by the median molecular length of the DNA (MML; determined electrophoretically; units = kilobase pairs).

[d]Mosquitofish were collected from a non-contaminated pond and caged in Pond 3513 for 2 weeks. RAPD genotypes were determined from DNA extracted from fin clips taken before and after placing fish in the cage. Data represent % survival of fish with and without certain RAPD bands.

[e]The name of the bands were derived from the name of the primer used to amplify them (designated by the manufacturer, e.g., UBC4, UBC12) followed by the molecular length (in base pairs) in subscript (e.g., $UBC4_{1610}$).

smallest genetic distance) to Crystal Springs is White Oak Lake. In fact, the Crystal Springs population is more genetically similar to both contaminated sites than it is to the other reference site, Wolf Creek. The fact that Crystal Springs is more closely related to Pond 3513 than to Wolf Creek is not surprising, given the fact that the Pond 3513 population originated from Crystal Springs. However, there is no a priori reason for suspecting such a close relationship between Crystal Springs and White Oak Lake. Interestingly, the degree of strand breakage in this population seems to be intermediate between the contaminated populations and Wolf Creek for liver DNA. When the single-strand MML is examined, Crystal Springs seems to have the same amount of damage as White Oak Lake (Fig. 1). This pattern is also reflected in the relationship between fecundity and genotype. For one contaminant-indicative band, there is a significantly higher fecundity for Crystal Springs fish with the band, as in the contaminated sites, whereas for another band, there is significantly higher fecundity for fish without it, as in Wolf Creek (Fig 3). Thus Crystal Springs may have genotoxic contamination not suspected before. Upon further investigation, it was subsequently found that Crystal Springs is contaminated with radon, on the order of 100 pCi/L (I.L. Larson, Oak Ridge National Laboratory, personal communication). It, therefore, appears that strand breakage and frequency of RAPD bands, and the specific relationship between genotype and fecundity, could be used to identify contamination where it was not previously suspected.

Besides the relationship between genotype and fecundity, the association of genotype and relative amount of DNA strand breakage is also a significant finding. First, although associations of genotype with fecundity and survival may provide evidence of contaminant-induced selection, they do not provide a possible mechanism whereby selection occurs. But the relationship between DNA damage and genotype suggests that individuals with the contaminant-indicative bands are better able to mitigate or repair the genotoxic effects of radioactive exposure than those without the band. One possibility is that these RAPD bands are amplified from or are genetically linked to radiation-responsive genes (e.g., antioxidant enzymes, DNA repair genes, stress proteins, etc.). There are also other genetic elements that have been shown to be involved in radiation response or radio-resistance, e.g., transposons (Dorling and Starlinger 1986) or DNA-nuclear matrix attachment regions (Schwartz and Vaughn 1989). An interesting observation is that both of these elements are bounded by inverted repeats, as are RAPD amplification sites, so the contaminant-indicative bands could be amplified from these elements as well. Identification of the sequence and location of RAPD amplification sites will need to be accomplished in order to confirm if any of these speculations are true.

Second, the relationship between strand breakage and genotype provides a direct link between population- and molecular-level responses to pollution. This is important because there have been a number of studies that have found differences in population genetic structure between contaminated and reference sites, but because there are many other variables which can affect population genetic structure, assigning a toxicological effect is difficult. However, if it can be established that there is a relationship between genotype and a response that is indicative of contaminant exposure (e.g., as DNA strand breakage is indicative of radiation exposure), this would provide evidence that these population differences are influenced by contamination exposure.

## Applicability

The above studies indicate that the RAPD technique can be used for examination of the effects of contaminant-induced selection on population genetic structure. It has been suggested previously that changes in population genetic structure may take place sooner than changes at the community or ecosystem levels of organization (Benton and Guttman 1992), proving useful as a population-level biomarker or an early warning indicator of such effects. In order for this to be true, a number of caveats must be satisfied. First, the organism that is monitored must respond relatively rapidly to selective pressures. Thus, the best choice for this type of biomonitoring would be an organism that is small, highly fecund, and has a short generation time. The various species of *Gambusia* meet these criteria (Tribault and Shultz 1978; Krumholz 1948). Second, the species must be genetically plastic, which means that they would be able to respond to selective pressures. This is typically found in species that are opportunistic generalists, such as members the *Gambusia* species complex. Third, changes in population genetic structure due to selection must be related to changes at higher levels of organization. As was discussed previously, the simplification of community structure that often occurs as a result of toxicant exposure is thought to result from adaptation of the generalist and not the specialist species, which are subsequently lost from the community (Luoma 1977). Therefore, evidence of genetic adaptation in generalized species may signal the potential for community-level changes. Finally, in order to be useful as an early warning indicator of higher-level effects, changes in population genetic structure as a result of selection must take place at contaminant concentrations below that at which community effects take place. This type of relationship has been found in several populations of small generalist fish species such as *G. affinis*, *Pimephales notatus*, and *F. heteroclitus* inhabiting metal contaminated environments in Iowa. The allelic frequencies of populations from contaminated habitats differed from those in reference populations. It was also found that fish with genotypes that were more prevalent in the contaminated populations were more resistant to metal toxicity. But in the field, there were no differences between reference and contaminated populations in terms of species diversity, richness, or evenness. This indicates that although the level of contamination was high enough to result in increases in metal-tolerant genotypes in small fish species, it was not high enough to cause community level changes overall (Roark and Brown 1996). Therefore, examination of population genetic structure and selective responses in *Gambusia* meets all of the aforementioned requirements.

The use of RAPDs in biomonitoring programs for this purpose may result in the reduction of cost and labor associated with ecological monitoring efforts. For example, determination of supraorganismal effects, such as calculation of indices of biotic integrity or community diversity are very time consuming and labor-intensive, often requiring the use of large boats, electroshocking equipment, crews of three or more people, and intensive sampling of organisms from multiple taxonomic levels. By contrast, collecting samples of small fish for RAPD analysis can be accomplished by two people with a seine, or even one person with a backpack shocker or dipnet. Also, once in the laboratory, every aspect of the assay from DNA extraction to agarose gel analysis can be automated, eliminating operator subjectivity and allowing low-cost analysis of large numbers of samples necessary for environmental monitoring. The most useful approach to using such information would be a tiered approach, in which the lower cost, less labor-intensive RAPD assay could be used to monitor large numbers of sites and could serve as a "red flag" to indicate in which areas to focus further ecological monitoring techniques. This approach would work best by comparing the similarity of population banding patterns between local populations, rather than on focusing on specific RAPD bands per se. Monitoring for specific RAPD bands would not be effective because RAPD bands are identified on a gel by molecular length, not by sequence. RAPD bands displayed by the same species of fish at two different geographic locations may be amplified from the same locus and may have homologous sequences, but they may be of different molecular length. Therefore, a more robust approach would be to determine the genetic distances between populations of interest. If one population is more different from the neighboring populations than would be expected based on population genetic theory, then this population would become the focus of more intensive investigation to determine if this disconcordance has an anthropogenic rather than a natural etiology. The more rigorous ecological sampling would then be applied only if an unexpected pattern emerges. This would be a much less cost- and labor-intensive approach than if all sites were continuously monitored for supraorganismal effects.

Another advantage of using the RAPD technique is that it is able to simultaneously amplify markers from large numbers of loci, increasing the possibility of finding useful polymorphisms. Other studies examining population genetic responses to pollution either 1) focus on "housekeeping" genes, and if a difference were found between contaminated and reference sites, try to explain why these genes would be responsive to contamination (e.g., Benton and Guttman 1991; Diamond et al. 1991), or 2) focus on genes that are responsive to contaminant exposure (e.g., cytochrome $P_{450}$ genes; Wirgin et al. 1991), then try to determine if there is a difference in the population genetics of these genes. Either way, this is a hit or miss strategy that may or may not yield answers and can result in wasted time and resources. The RAPD approach is unique in that it examines a large number of loci. Then, once differences are found between contaminated and reference sites, those markers that show such differences can be isolated and identified. Because more loci are simultaneously examined, this increases the possibility of finding useful polymorphisms. This approach focuses on finding differences and then identifying the genes responsible, as opposed to identifying genes and then trying to determine if population differences exist. The genomic locus from which the RAPD bands are amplified can be isolated and/or identified. Such information could lead to development of new biomarkers for monitoring individual-level effects and identification of novel genetic markers for monitoring population-level effects. For example, the gene or genetic element from which the RAPD band is amplified could be isolated and cloned from genomic libraries and used as a probe for population genetic studies (e.g., genetic diversity or frequency of specific haplotypes). Use of these genomic loci for biomonitoring would be a more valid strategy than monitoring for specific RAPD bands because the loci from which the RAPD bands are amplified would be more likely to be conserved than the presence of the band itself.

The use of RAPD markers to identify contaminant-responsive loci could further the understanding of the genetic basis of xenobiotic resistance. This can be used to compliment and elucidate results from other types of biomonitoring, such as examination of biomarker responses, population/community responses, or toxicity tests. In some polluted sites, certain biomarker responses are elevated relative to control sites, but are not correlated with effects at the population or community level (Schlenk et al. 1996). In some situations, a site may be heavily polluted, but the biomarker responses may not be much different from those from a reference site (Landner et al. 1994.). If a lack of such relationships is due to the evolution of xenobiotic resistance in the exposed populations, this could be addressed by examination of population genetic structure, particularly if pollutant-resistant genotypes are identified.

Effective biomonitoring requires the integration of both laboratory and field studies, but there can be discrepancies between results from the laboratory and the field. For instance, patterns of biomarker response in the laboratory may be different than those in the field (Theodorakis et al. 1992), and toxicity tests in the laboratory may not accurately predict toxic affects on natural populations (van Loon and Beamish 1977; Snell et al., chapter 9; Baird and Barata, chapter 11). One contributing factor may be the development of resistance in the field populations. Hence, the use of RAPDs to investigate the effects of xenobiotic selection at polluted sites could be used to elucidate such discrepancies and may improve methods of extrapolation of toxicity tests to field situations.

Understanding the genetic basis of xenobiotic resistance and the effects of pollutant-induced selection are also applicable to the assessment of human health risks. It has been found that susceptibility to genotoxicants or other xenobiotic stressors is variable and genetically determined in human populations (Rudiger 1991; Nebert et al. 1991). Consequently, when humans are exposed to environmental pollutants, there may be selective processes occurring in the form of differential mortality, carcinogenesis, teratogenesis, or fertility. Effects of selection on natural populations may give an indication of possible effects on sensitive genotypes in the human population. Second, understanding the molecular basis of genotoxicant-induced selection and adaptation in natural populations can lead to a better understanding of the genetic basis of differential susceptibility to genotoxicant exposure and its associated diseases (e.g., cancer and teratogenesis) in humans.

## CONCLUSIONS

Determination of pollutant-caused selection can be an important component of ecological risk assessment. First, failure to account for such effects may lead to biased risk estimates. Second, adaptation to pollutants may lead to reduction in fitness of individuals and viability of populations. Third, evidence of selection in generalist species may signal the potential of changes in community structure.

The results from the present study indicate that radionuclide-induced selection is able to modify population genetic structure in *G. affinis*, as indicated by using the RAPD technique. Evidence that selection has occurred comes from integration of results from RAPD band frequencies and genotype-specific differences in fecundity, survival, and DNA damage. Comparisons of genetic differences between localized populations or the use of RAPD-derived markers may be widely applicable to monitoring the effects of a large number of pollutants and a diverse array of species. RAPD genotype analysis may be a way to relate molecular-level effects with those at higher levels of organization, and use of this technique seems promising for the development of population-level biomarkers, which are early-warning indicators of supraorganismal effects.

## FUTURE DIRECTIONS

Although the present study provides compelling evidence that RAPD bands may be markers of contaminant-responsive loci, definitive evidence will not be available until the molecular identity of the bands or their respective amplification sites are known. Therefore, future efforts will be directed toward this goal via cloning and sequencing; Southern, Northern and dot-blotting; and exploration and identification of differentially expressed mRNA species from radiation-exposed fish and their relation to RAPD genotype expression. Also, the present results deal only with radionuclide-exposed *G. affinis* from Oak Ridge, TN. Further efforts will determine the applicability of this technique to other species exposed to different toxicants and from a variety of geographic localities. Finally, efforts are underway in order to determine specific relationships between RAPD population genetic structure and alterations in population, community, and ecosystem structure and function at polluted sites.

## REFERENCES

Ahmad, S. 1995. Oxidative stress from environmental pollutants. *Archives of Insect Biochemistry and Physiology* 20:145–157.

Albert, P. L. 1993. Strategies for population reintroduction: Effects of genetic variability on population growth and size. *Conservation Biology* 7:194–199.

Allen, W. R., and P. M. Shepard. 1971. Copper tolerance in some California populations of the monkey flower *Mimmulus guttatus*. *Proceedings of the Royal Society of London B* 177:177–196.

Baturo, W., and L. Lagadic. 1996. Benzo(a) pyrene hydroxylase and glutathione S-transferase activities as biomarkers in *Lymnea plaustris* (Mollusca, Gastropoda) exposed to atrazine and hexachlorobenzene in freshwater mesocosms. *Environmental Toxicology and Chemistry* 15:771–781.

Benton, M. J., and S. I. Guttman. 1992. Allozyme genotype and differential resistance to mercury pollution in the caddisfly, *Nectopsyche albida*. I. Single-locus genotypes. *Canadian Journal of Fisheries and Aquatic Sciences* 49:142–146.

Bickham, J. W., and M. Smolen. 1994. Somatic and heritable effects of environmental genotoxins and the emergence of evolutionary toxicology. *Environmental Health Perspectives* 102:25–28.

Black, M. C., J. R. Ferrell, R. C. Homing, and L. K. Martin, Jr. 1996. DNA strand breakage in freshwater mussels (*Anodonta granis*) exposed to lead in the laboratory and field. *Environmental Toxicology and Chemistry* 15:802–808.

Blank, H., and S.-A. Wängberg. 1991. Patterns of cotolerance in marine periphyton communities established under arsenate stress. *Aquatic Toxicology* 21:1–14.

Blaylock, B. G. 1969. The fecundity of a *Gambusia affinis affinis* population exposed to chronic environmental radiation. *Radiation Research* 37:108–117.

Blaylock, B. G., M. L. Frank, F. O. Hoffman, L. A. Hook, G. W. Suter, and J. A. Watts. 1991. *Screening of contaminants in Waste Area Grouping 2 at Oak Ridge National Laboratory, Oak Ridge, Tennessee*. Oak Ridge, TN: Oak Ridge National Laboratory.

Bradley, B. P., C. M. Gonzales, J. A. Bond, and B. E. Tepper. 1994. Complex mixture analysis using protein expression as a qualitative and quantitative tool. *Environmental Toxicology and Chemistry* 13:1043–1050.

Bucheli, T. D., and K. Fent. 1995. Induction of cytochrome P450 as a biomarker for environmental contamination in aquatic systems. *Critical Reviews in Environmental Science and Technology* 25:201–268.

Changon, N. L., and S. I. Guttman. 1989a. Differential survivorship of allozyme genotypes in mosquitofish populations exposed to copper or cadmium. *Environmental Toxicology and Chemistry* 8:319–326.

Changon, N. L., and S. I. Guttman. 1989b. Biochemical analysis of allozyme copper and cadmium tolerance in fish using starch gel electrophoresis. *Environmental Toxicology and Chemistry* 8:1141–1147.

Cox, R. M., and T. C. Hutchinson. 1979. Metal co-tolerances in the grass *Deschampa cespitosa*. *Nature* 279:231–233.

Depledge, M. H. 1996. Genetic ecotoxicology—an overview. *Journal of Experimental Marine Biology and Ecology* 200:57–66.

Diamond, S. A., M. C. Newman, M. Mulvey, and S. I. Guttman. 1991. Allozyme genotype and time-to-death of mosquitofish, *Gambusia holbrooki*, during acute inorganic mercury exposure: A comparison of populations. *Aquatic Toxicology* 21:119–134.

Dinesh, K. R., W. K. Chan, T. M. Lim, and V. P. E. Phang. 1993a. RAPD markers in fishes: An evaluation of resolution and reproducibility. *Asia Pacific Journal of Molecular Biology and Biotechnology* 3:112–118.

Dinesh, K. R., T. M. Lim, K. L. Chua, W. K. Chan, and V. P. E. Phang. 1993b. RAPD analysis: An efficient method of DNA fingerprinting in fishes. *Zoological Science* 10:849–854.

Dorling, H. P., and J. Starlinger. 1986. Molecular genetics of transposable elements in plants. *Annual Review of Genetics* 20:175–200.

Duncan, D. A., and J. F. Klaverkamp. 1983. Tolerance and resistance in white suckers (*Catostomus commersonii*) previously exposed to cadmium, mercury, zinc, or selenium. *Canadian Journal of Fisheries and Aquatic Sciences* 40:128–138.

Erickson, J. M., M. Rahire, J.-D. Rochaix, and L. Mets. 1985. Herbicide resistance and cross-resistance: Changes at three distinct sites in the herbicide-binding protein. *Science* 228:204–207.

Gillespie, R. B., and S. I. Guttman. 1988. Effects of contaminants on the frequencies of allozymes in populations of central stonerollers. *Environmental Toxicology and Chemistry* 8:309–317.

Guttman, S. I. 1994. Population genetic structure and ecotoxicology. *Environmental Health Perspectives* 102(Suppl 12):97–100.

Hadrys, H., M. Balick, and B. Schierwater. 1992. Applications of random amplified polymorphic DNA (RAPD) in molecular ecology. *Molecular Ecology* 1:55–63.

Hoare, K., A. R. Beaumont, and J. Davenport. 1995. Variation among populations in the resistance of *Mytilus edulis* embryos to copper: Adaptation to pollution? *Marine Ecology Progress Series* 120:155–161.

Hubert, P. D. N., and M. M. Luiker. 1996. Genetic effects of contaminant exposure—towards an assessment of impacts on animal populations. *Science of the Total Environment* 191:23–58.

Kirsch, R. F., M. B. Flick, and C. N. Trumbore. 1991. Radiation chemical mechanisms of single and double-strand break formation in irradiated SV 40 DNA. *Radiation Research* 126:251–259.

Klerks, P. L., and J. S. Levinton. 1989. Rapid evolution of metal resistance in a benthic oligocheate inhabiting a metal-polluted site. *Biological Bulletin* 176:135–141.

Klerks, P. L., and P. R. Bartholomew. 1991. Cadmium accumulation and detoxification in a Cd-resistant population of the oligochaete *Limnodrilus hoffmeisteri*. *Aquatic Toxicology* 19:97–112.

Krebs, C. J. 1989. *Ecological methodology*. New York: Harper and Collins Publishers.

Krumholz, L. A. 1948. Reproduction in the western mosquitofish *Gambusia affinis* and its use in mosquito control. *Ecological Monographs* 18:1–43.

Kubitz, J. A., E. C. Lewek, J. M. Besser, J. B. Drake, III, and J. P. Geisy. 1995. Effects of copper-contaminated sediments on *Hyella azteca, Daphnia magna*, and *Ceriodaphnia dubia*: Survival, growth, and enzyme inhibition. *Archives of Environmental Contamination and Toxicology* 29:97–103.

Kuusipalo, L. 1994. *Assessing genetic structure of pelagic fish populations of Lake Tanganyika*. (In French). Bujumbura, Burundi: Bujumbura-Burundi FAA.

Land, R., and S. Shannon. 1996. The role of genetic variation in adaptation and population persistence in a changing environment. *Evolution* 50:434–437.

Landner, L., O. Grahn, J. Hardig, K.-J. Lehtinen, C. Monfelt, and J. Tana. 1994. A field study of environmental impacts at a bleached kraft pulp mill site on the Baltic Sea coast. *Ecotoxicology and Environmental Safety* 27:128–157.

Lee, C. J., M. C. Newman, and M. Mulvey. 1992. Time to death of mosquitofish (*Gambusia holbrooki*) during acute inorganic mercury exposure: Population structure effects. *Archives of Environmental Contaminations and Toxicology* 22:284–287.

Luoma, S. N. 1977. Detection of trace contaminant effects on aquatic ecosystems. *Journal of the Fisheries Research Board of Canada* 34:436–439.

Lynch, M., and B. G. Milligan. 1994. Analysis of population genetic structure with RAPD markers. *Molecular Ecology* 3:91–99.

Macnair, M. R., S. E. Smith, and Q. J. Cumbes. 1993. Heritability and distribution of variation in degree of copper tolerance in *Mimulus guttatus* at Copperopolis, California. *Heredity* 71:445–455.

Michelmore, R. W., I. Paran, and R. V. Desseli. 1991. Identification of markers linked to disease-resistance genes by bulked segregant analysis: A rapid method to detect markers in specific genomic regions by using segregating populations. *Proceedings of the National Academy of Science* 88:9828–9832.

Mulvey, M., M. C. Newman, A. Chazal, M. M. Keklak, M. G. Heagler, and L. S. Hales, Jr. 1995. Genetic and demographic responses of mosquitofish (*Gambusia holbrooki* Girard 1859) populations stressed by mercury. *Environmental Toxicology and Chemistry* 14:1411–1418.

Naish, K. A., M. Warren, F. Bardakci, D. O. F. Skibinski, G. R. Carvalho, and C. G. Mair. 1995. Multilocus DNA fingerprinting and RAPD reveal similar genetic relationships between strains of *Oreochromis niloticus* (Pisces: Cichlidae). *Molecular Ecology* 4:271–274.

Nebert, D. W., D. D. Peterson, and A. Puga. 1991. Human AH locus polymorphism and cancer: Inducibility of CYP1A1 and other genes by combustion products and dioxin. *Pharmacogenetics* 1:68–78.

Newman, M. C., S. A. Diamond, M. Mulvey, and P. Dixon. 1989. Allozyme genotype and time to death of mosquitofish, *Gambusia affinis* (Baird and Girard) during acute toxicant exposure: A comparison of arsenate and inorganic mercury. *Aquatic Toxicology* 15:141–156.

Olive, P. L. 1992. DNA organization affects cellular radiosensitivity and detection of initial DNA strand breaks. *International Journal of Radiation Biology* 62:389–396.

Postlethwait, J. H., S. L. Johnson, C. N. Midson, W. S. Talbot, and M. Gates. 1994. A genetic linkage map for the zebrafish. *Science* 264:699–703.

Posthuma, L., R. F. Hogervorst, and N. M. van Straalen. 1992. Adaptation to soil pollution by cadmium excretion in natural populations of *Orchesella cincta* (L.) (Collembola). *Archives of Environmental Contamination and Toxicology* 22:146–156.

Postma, J. F., A. Vankleunen, and W. Admiraal. 1995. Alterations in life-history traits of *Chironomus riparius* (Diptera) obtained from metal contaminated rivers. *Archives of Environmental Contamination and Toxicology* 29:469–475.

Prince, R., and K. R. Cooper. 1995. Comparisons of the effects of 2,3,7,8 tetrachlorodibenzo-p-dioxin on chemically impacted and nonimpacted subpopulations of *Fundulus heteroclitus*: II. Metabolic considerations. *Environmental Toxicology and Chemistry* 14:589–595.

Roark, S., and K. Brown. 1996. Effects of metal contamination from mine tailings on allozyme distributions of populations of great plains fishes. *Environmental Toxicology and Chemistry* 15:921–927.

Roesijadi, G. 1992. Metallothioneins in metal regulation and toxicity in aquatic animals. *Aquatic Toxicology* 22:81–114.

Roots, R., and S. Okada. 1975. Estimation of life times and diffusion distances of radicals in X-ray induced DNA strand breaks and killing of mammalian cells. *Radiation Research* 64:306–320.

Rudiger, H. W. 1991. Genetic susceptibility to DNA methylating agents. In *Ecogenetics*, ed. P. Grandjean, 205–215. London: Chapman and Hall.

Sanders, B. M. 1996. Stress proteins in aquatic organisms: an environmental perspective. *Critical Reviews in Toxicology* 23:49–75.

Schlenk, D., E. J. Perkins, G. Hamilton, Y. S. Zhang, and W. Layher. 1996. Correlation of hepatic biomarkers with whole animal and population-community metrics. *Canadian Journal of Fisheries and Aquatic Sciences* 53:2299–2309.

Schuleter, M. A., S. I. Guttman, J. T. Oris, and A. J. Bailer. 1997. Differential survival of fathead minnows, *Pimephales promelas*, as affected by copper exposure, prior population stress, and allozyme genotypes. *Environmental Toxicology and Chemistry* 16:939–947.

Schwartz, J. L., and T. M. Vaughn. 1989. Association among DNA/chromosome break rejoining rates, chromatin structure, and radiation sensitivity in human tumor cell lines. *Cancer Research* 49:5054–5057.

Shugart, L. R. 1993. Genotoxic responses in blood. In *Nondestructive biomarkers in vertebrates*, eds. M. C. Fossi and C. Leonzio, 131–145. Boca Raton, FL: Lewis Publishers, Inc.

Shugart, L. R., and C. W. Theodorakis. 1994. Environmental genotoxicity: Probing the underlying mechanisms. *Environmental Health Perspectives* 102(Suppl 12):13–17.

Shulte-Frohlinde, D., and C. von Sonntag. 1985. Radiolysis of DNA and model systems in the presence of oxygen. In *Oxidative stress*, ed. H. Seiss, 11–40. London: Academic Press.

Sulatos, L. G. 1994. Mammalian toxicology of organophosphrous pesticides. *Journal of Toxicology and Environmental Health* 43:271–289.

Tamura, T., O. M. Sealand, and J. O. Duguid. 1977. *Preliminary inventory of $^{239,240}Pu$, $^{90}Sr$, and $^{137}Cs$ in waste pond No. 2 (3513)*. Oak Ridge, TN: Oak Ridge National Laboratory.

Theodorakis, C. W., B. G. Blaylock, and L. R. Shugart. 1997. Genetic ecotoxicology: I. DNA integrity and reproduction in mosquitofish exposed in situ to radionuclides. *Ecotoxicology* 5:1–14.

Theodorakis, C. W., S. J. D'Surney, J. W. Bickham, T. B. Lyne, B. P. Bradley, W. E. Hawkins, W. L. Farkas, J. F. McCarthy, and L. R. Shugart. 1992. Sequential expression of biomarkers in bluegill sunfish exposed to contaminated sediment. *Ecotoxicology* 1:45–73.

Theodorakis, C. W., and L. R. Shugart. 1997. Genetic ecotoxicology: II. Population genetic structure in radionuclide-contaminated mosquitofish (*Gambusia affinis*). *Ecotoxicology* 6:335–354.

Theodorakis, C. W., and L. R. Shugart. 1998a. Genetic ecotoxicology: III. Relationship between genotype and DNA integrity in mosquitofish exposed to radionuclides. *Ecotoxicology* 7: (in press).

Theodorakis, C. W., and L. R. Shugart. 1998b. Genotype-dependant survivorship and DNA integrity of mosquitofish exposed to radionuclides. *Environmental Toxicology and Chemistry* (in press).

Trabalka, J. R., and C. P. Allen. 1977. Aspects of fitness of a mosquitofish *Gambusia affinis* exposed to chronic low-level environmental radiation. *Radiation Research* 70:198–211.

Tribault, R. E., and R. J. Shultz. 1978. Reproductive adaptations among viviparous fishes (Poeciliidae). *Evolution* 32:320–333.

Urwin, P. E., Q. J. Groom, and N. J. Robinson. 1996. Characterization of two cDNAs and identification of two proteins that accumulate in response to cadmium in cadmium-tolerant *Datura innoxia* cells. *Journal of Experimental Botany* 47:1019–1024.

van Loon, J. C., and R. J. Beamish. 1977. Heavy metal contamination by atmospheric fallout of several Flin Flon area lakes and the relations to fish populations. *Journal of the Fisheries Research Board of Canada* 34:899–906.

Weis, J. S., P. Weis, and M. Heber. 1982. Variation in response to methylmercury by killifish (*Fundulus heteroclitus*) embryos. In *Aquatic toxicology and risk assessment: 5th conference (ASTM S.T.P. 776)*, eds. J. G. Pearson, R. Foster, and W. E. Bishop. Philadelphia: American Society for Testing and Materials.

Welsh, J. N., and D. M. McClelland. 1990. Fingerprinting genomes using PCR with arbitrary primers. *Nucleic Acids Research* 18:7213–7218.

Whitton, B. A., and F. H. A. Shehata. 1982. Influence of cobalt, nickel, copper, and cadmium on the blue-green alga *Anacystis nidulans*. *Environmental Pollution Series A* 27:275–281.

Williams, J. G. K., A. R. Kubelik, K. J. Livak, J. A. Rafaski, and S. V. Tingey. 1990. DNA polymorphisms amplified by arbitrary primers are useful as genetic markers. *Nucleic Acids Research* 18:6531–6535.

Wirgin, I., G.-L. Kreamer, and S. J. Garte. 1991. Genetic polymorphism of cytochrome P-450IA in cancer-prone Hudson River tomcod. *Aquatic Toxicology* 19:205–214.

Woodhead, D. S. 1977. The effects of chronic irradiation on the breeding performance of the guppy (*Poecilia reticulata*) (Osteichthes: Teleosti). *International Journal of Radiation Biology* 32:1–22.

*Genetics and Ecotoxicology*
Edited by V. E. Forbes
Copyright © 1999 Taylor & Francis

# 8

# Structural Damage to DNA in Response to Toxicant Exposure

## Lee R. Shugart

**Abstract.** Substances capable of modifying the genetic material of living organisms are termed genotoxicants. Many of these are anthropogenic chemicals, and although widely distributed in our environment, their long-term ecological effects are largely unknown. Genotoxicants can interact with DNA and potentiate structural alterations in this important cellular macromolecule. The number of structural alterations per genome is often quite small and therefore sensitive analytical methods are required to detect the unique types that occur. Several of the current methodologies for the detection of changes, alterations, and modifications to the structure of DNA are reviewed. Specific types of damage that occur to the structure of DNA can be used as endpoints for assessing exposure to genotoxicants. Persistence of damage may potentiate deleterious responses at other levels of biological organization which, in turn, become endpoints of health-related effects. The relevance of genetic damage for the prediction of effects at the population, community, and ecosystem levels is more problematic.

**Keywords.** DNA adducts; DNA repair; DNA strand breakage; genotoxicants; molecular genetics.

## INTRODUCTION

Chemicals and/or physical agents are considered genotoxicants if they modify the structure of DNA, the genetic moiety of living organisms, and adversely affect its integrity (structure and/or function). Genetic toxicology are an area of science in which the interactions of genotoxicants with the cell's genetic material is studied in relation to subsequent effects(s) on the health of the organism. The occurrence in the environment of genotoxicants of anthropogenic origin is one of those complex toxicological problems that have become a major concern of society. This concern arises in part because some are known carcinogens and mutagens.

A variety of approaches have been employed by ecotoxicologists to detect both direct and inferred changes to the structure of DNA, including assays that demonstrate nucleotide base modification (Shugart et al. 1987, 1990c; Dunn et al. 1987; Malins and Haimanot 1991; McCarthy et al. 1991), DNA strand breakage (Shugart 1988, 1996b), sister chromatid exchange or other chromosomal aberrations (Dixon 1982; Maddock et al. 1986), unscheduled DNA synthesis (West et al. 1988), micronuclei formation (Das and Nanda 1986; Hose et al., 1987), point mutations (Yung-Jin et al.

1991; McMahon et al. 1990), and variations in nuclear DNA content (Bickham and Smolen, 1994). The consequences from exposure to genotoxicants may pose serious biological problems because of the potential for these substances to produce adverse change at the cellular and organismal levels as well as to the offspring of the exposed organisms (Wurgler and Kramers 1992; Shugart and Theodorakis 1996).

The focus of this chapter is on the various types of DNA structural damage that exposure to genotoxicants may elicit. Particular emphasis is placed on available techniques and methodologies with the sensitivity and specificity to accomplish the task of identification. Examples of the use of DNA structural damage to assess environmental problems are listed. Approaches and investigations that evaluate and document the ecological consequences of exposure of populations in the field to genotoxicants are discussed in more detail in other chapters in this book. The reader is referred to chapter 7 by Theodorakis and Shugart on natural selection in contaminated environments.

## DEOXYRIBONUCLEIC ACID

Figure 1 is a schematic representation of the status of cellular DNA in relation to various processes that disrupt DNA integrity and those that maintain DNA integrity (Shugart 1990a; Shugart et al. 1992). It is intended as a focal point for and to simplify the discussion of DNA-genotoxicant interactions, which are covered in more detail below.

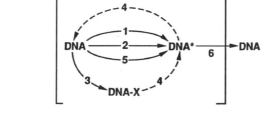

| INSULT | REPAIR | SYNTHESIS |
|---|---|---|
| 1. Normal wear and tear | 4. Incision | 5. Replication |
| 2. UV and ionizing radiation | excision | 6. Postreplication |
| ($\gamma$ and X irradiation) | resynthesis | modification |
| 3. Chemical | ligation | |

DNA:     Normal double-stranded DNA with no strand breaks.
DNA-X:  Chemically modified DNA.
DNA*:   DNA with strand breaks.

**FIG. 1.** A schematic representation of the status of DNA in relation to insults that disrupt DNA integrity and cellular processes that maintain DNA integrity (reprinted from Shugart 1990a, with permission of Lewis Publishers, Boca Raton, FL).

## DNA Integrity

DNA is usually present in the nucleus of living cells as a functionally stable, double-stranded entity without discontinuity (strand breaks) or abnormal structural modifications (adducts or chemically modified, missing, or added bases) and exists as chromatin, a DNA-protein complex. In this state, it is considered to have a high degree of integrity. Many physiological processes temporally change and disrupt the integrity of DNA (e.g., replication, gene expression, post translational modification) but are accompanied by elaborate enzymatic processes that maintain and return the DNA to its state of high integrity (Elliott and Elliott 1997). The rigid maintenance of this integrity is important for survival and is reflected in the low mutational rate observed in living organisms.

## Pathways Leading to Structural Modifications of DNA

Normal cellular events can influence the structural integrity of cellular DNA in a defined, programmed manner, and those modifications and changes that result usually exist for a short period of time. For example, during the cellular process of replication (Fig. 1, pathway 5), the prevention of supercoiling involves the transient breakage of the polynucleotide chain. Thus, temporal modifications in the form of DNA strand breaks occur (Fig. 1, DNA*). Other normal structural modifications may persist for longer periods of time, such as the methylation of specific DNA bases, a post-translational modification of DNA (Fig. 1, pathway 6).

Structural modifications can occur to DNA by other than normal processes and result in damaged DNA, which, if left uncorrected, may potentiate subsequent problems for the cell. Strand breaks in DNA (Fig. 1, DNA*) can occur via apparently innocuous events such as metabolism and random thermal collisions by cellular molecules (Fig. 1, pathway 1). Exposure to genotoxicants such as physical agents, e.g., ultraviolet light and ionizing radiation (Fig. 1, pathway 2) and certain chemicals (Fig. 1, pathway 3) can produce structural modifications ranging from adducts (Fig. 1, DNA-X) to strand breaks. Some chemical and physical agents cause strand breaks directly via cellular mechanisms that generate free-radicals, whereas other genotoxicants interfere with the fidelity of DNA repair (Fig. 1, pathway 4) or prevent normal modification of the DNA molecule (Fig. 1, pathway 6). Damage such as the loss of bases from the DNA molecule frequently occurs as a result of random thermal collisions (Fig. 1, pathway 1) or from the breakdown of chemically unstable adducts.

Thus, at any one point in time there may be present in the cell of an organism a DNA molecule with a variety of structural modifications. Fortunately, most cells have the enzymatic capacity to repair structural modifications in DNA (Elliott and Elliott 1997) which, under normal circumstances, efficiently eliminates the damage (Fig. 1, pathway 4).

## DETECTION OF STRUCTURAL DAMAGE TO DNA

Early attempts at assessing the effects of genotoxicants on natural field populations most often involved direct observation such as the visual occurrence of neoplasms and chromosomal aberrations in various plants, wild terrestrial mammals, and aquatic vertebrates (Sandhu and Lower 1989). The development of neoplasia, tumors, or other pathological responses characteristic of carcinogenesis is a slow process (months to years) that is dependent upon numerous unknown and ill-defined cellular processes (Thilly and Call 1986). These endpoints, although useful for defining the health of the organism, are less useful when attempting to identify the causative agent. With respect to genotoxicants, this limitation may be circumvented by focusing on specific cellular events that occur to DNA shortly after exposure.

Exposure of an organism to genotoxicants may cause the induction of a cascade of cellular events that results in (or is closely associated with) structural modifications to the DNA (Shugart 1996a). Various types of structural alterations are possible (e.g., adducts, altered bases, strand breaks, and changes in minor nucleotide content). Many factors can contribute to the final observable damage within the DNA. Important among these are the chemical and/or physical properties of the genotoxicant, the species of organism exposed, the route of exposure, as well as the life stage of the organism and its genetic makeup. As previously noted, changes to the structure of DNA are usually amenable to correction by the cell's DNA repair machinery (Fig. 1, pathway 4). However, damage that is not corrected or is improperly processed may potentiate irreversible cellular events that result in the appearance, after cell division, of abnormally processed DNA (e.g., chromosomal aberrations, micronuclei, etc.).

The temporal expression of the types of biological responses that can be observed upon exposure to genotoxicants are summarized in Table 1. Two points need to be emphasized. First, these responses are points on a continuum and are subject to change and redefinition as our knowledge in genetic toxicology advances, but they can be used

**TABLE 1.** *Cellular responses of exposure to genotoxicants*

| Biological response | Expression in cell | Temporal occurrence[a] |
|---|---|---|
| Detoxication | Induction of P450 enzyme system | Early |
| DNA adduct | Covalent attachment of genotoxicant to DNA | Early |
| DNA strand breaks | Breakage of DNA phosphodiester linkages | Early |
| DNA repair | Induction of DNA repair enzymes | Early |
| DNA base modification | Hypomethylation of 5-methylcytosince bases | Early/Middle |
| Abnormal DNA | Chromosomal abberations, micronuclei, aneuploidy | Middle/Late |

[a]Temporal occurrence subsequent to exposure will depend on species and type of genotoxicant. Early: hours to days; Middle: days to weeks/months; late: weeks/months to years.

as markers to test for genotoxicity (Shugart 1990a, 1995; Shugart et al. 1992). The identification of a specific DNA adduct, for example, represents unequivocally the exposure of an organism to a specific genotoxicant. Second, because genotoxicants are usually present in the environment at low concentrations, organismal exposure results in very low levels of modified or damaged DNA. Therefore, special analytical methods with high sensitivity are needed. Methods that measure these biological markers of genotoxicity, especially structural damage to DNA, are discussed below in relation to their use in environmental studies.

Adducts and strand breaks are the two structural modifications to DNA that have been studied extensively. However, this would not have occurred if it weren't for the availability of analytical techniques with the appropriate selectivity and sensitivity. In this context it should be noted that the success of current in situ efforts in the field of environmental genotoxicology (Shugart 1995) is due mainly to the use of modern analytical techniques currently available in the fields of chemistry and molecular biology. These analytical capabilities have in turn advanced our knowledge about the cellular mechanisms involved in the genotoxic response.

## DNA Adducts

### *Introduction*

The adverse health effects of most environmental chemicals are the result of their covalent binding to physiologically important receptor molecules. Organophosphate poisoning and chemically-induced cancer are examples of toxic responses. Both occur as the result of the covalent binding of a specific chemical to a receptor molecule. With the former it is to a protein (cholinesterase) and with the latter to DNA. Identification of the interactive products can represent not only the most direct and biologically relevant indicator of exposure to a toxicant, but also the risk from that exposure.

The majority of chemicals that are genotoxicants exert their effects only after metabolic conversion to chemically reactive forms (Phillips and Sims 1979; Harvey 1982; Shugart 1996a). Under certain conditions and depending upon the chemical properties of the genotoxicant, the reactive form of the genotoxicant may bind covalently to cellular macromolecules, including nucleic acids and proteins. The product of this reaction with DNA is termed a DNA adduct (Fig. 1, pathway 3, DNA-X). Some genotoxicants are so chemically reactive that metabolic activation is not necessary and direct adduction to cellular molecules occurs without metabolic assistance.

Currently, methods of varying sensitivity exist to detect and quantify DNA adducts (Marnett 1993; Qu et al. 1997). The most prominent is $^{32}$P-postlabeling, but others, including high performance liquid chromatography (HPLC/fluorescence spectrophotometry) and immunoassays using adduct-specific antibodies, have found application. The methods discussed below have been detailed elsewhere (Shugart et al. 1992; Shugart 1994a, 1994b, 1995, 1996a).

**TABLE 2.** *Environmental monitoring: detection of adducts by [32] P-postlabeling*

| Environment/Contaminant | Species | Reference |
|---|---|---|
| Freshwater Stream (US)/ sediment-bound PAHs | Catfish | Dunn et al. 1987 |
| Freshwater Stream (Europe)/ complex industrial waste | Various Fish Species | Kurelec et al. 1989 |
| Marine (Adriatic)/complex industrial waste | Mussel<br>Carp | Kurelec et al. 1990<br>Kurelec et al. 1992 |
| Marine Harbor (US)/PAHs & complex industrial waste | English Sole &<br>  Winter Flounder | Varanasi et al. 1989<br>French et al. 1996 |
| Marine/Arctic (Canada)/organics and PAHs | Beluga Whale | Ray et al. 1991 |
| Estuarine River (US)/complex industrial waste | Muskrats | Halbrook et al. 1992 |
| Marine Harbor (Europe)/PAHs & complex industrial waste | Eel | van Schooten et al. 1995 |
| Estuarine River (Canada)/complex industrial waste | White Sucker Fish | El-Adlouni et al. 1995 |
| Terrestrial/ PAH-contaminated soil | Earth Worm | Walsh et al. 1995 |

## [32]P-Postlabeling

Many lines of evidence implicate the DNA adduct as a key element in the initiation of chemical carcinogenesis (Weinstein 1978; Wogan and Gorelick 1985). As a result, the [32]P-postlabeling technique was originally developed as a general method for studying the binding of known carcinogens to DNA (Randerath et al. 1981). The technique has been well documented for the measurement of extremely low levels of DNA adducts in laboratory animals and human tissue (Bartsch et al. 1988). Currently this methodology is finding application in environmental monitoring studies for the detection of adducted chemicals to DNA in numerous species suspected of exposure to genotoxicants (Qu et al. 1997; see Table 2).

The methodology used for the [32]P-postlabeling adduct assay consists of the following steps. DNA isolated from a suitable sample is enzymatically hydrolyzed with micrococcal nuclease and spleen phosphodiesterase. This results in the formation of $3'$-monophosphates of both normal and adducted nucleosides. The products are then [32]P-labeled by T4 polynucleotide kinase-catalyzed phosphorylation with $[\gamma\text{-}^{32}P]ATP$, leading to $5'\text{-}^{32}P$-labeled $3',5'$-bisphosphate nucleosides.

[32]P-labeled adducts are resolved from normal nucleotides, $^{32}Pi$, unused $[\gamma\text{-}^{32}P]ATP$, and unknown contaminants by anion exchange TLC on polyethyleneimine-cellulose using a multidirectional developmental scheme (Gupta et al. 1982; Gupta and Randerath 1988). Dried chromatograms are placed in contact with X-ray film and exposed at $-80°C$. Adducts are detected by autoradiography and quantified by scintillation counting. Genotoxicant-related adduct patterns are obtained from genotoxicant-exposed DNA.

The sensitivity of the assay for adducts can be enhanced by several orders of magnitude and is most often accomplished via the pre-enrichment of aromatic/hydrophobic nucleosides before labeling. Extracting in 1-butanol in the presence of a phase-transfer agent is one method (Gupta 1985). An alternative procedure (Reddy and Randerath 1986) involves selective enzyme hydrolysis of the 3′nucleotides with nuclease P1, or combined with reversed phase HPLC (Dunn and San 1988).

Detection of adduct levels as low as one in $10^{10}$ normal nucleotides is possible by this technique (Qu et al. 1997); however, $^{32}$P-labeled spots of undetermined origin also occur and may interfere with the interpretation of adduct patterns (Randerath et al. 1986; Kurelec et al. 1989). Nevertheless, with the exception of methylating agents and mycotoxins, numerous known carcinogens have been shown to produce carcinogen-specific fingerprints by this method (Gupta and Randerath 1988).

### *HPLC/Fluorescence*

An alternative method to the $^{32}$P-postlabeling technique is detection of the binding of fluorescent chemicals to DNA (Rahn et al. 1982). This technique involves removal of the adduct from DNA (usually by acid hydrolysis) and separation by high performance liquid chromatography (HPLC) coupled with fluorescence analysis. Application of this technique is limited by the presence and release of an intact fluorescent moiety. Shugart et al. (1983) have shown that adduct formation at the femtomole level between the ubiquitous chemical carcinogen, benzo[a]pyrene, and DNA can be detected and quantified by this method in both mice and fish exposed under laboratory conditions (Shugart and Kao 1985; Shugart et al. 1987). The technique has also been used in environmental studies to detect benzo[a]pyrene adducts in the DNA of beluga whales from the St. Lawrence River in Canada (Martineau et al. 1988).

### *Immunological Techniques*

Antibodies developed against DNA adducts produced by a specific genotoxicant have been used to examine DNA for antigenic properties. A major limitation of this approach is the need to develop specific antibodies and establish the specificity of the antibody for each genotoxicant-induced DNA adduct or class of adducts of interest. This is an important consideration when exposure is to a complex mixture of environmental chemicals. In addition, sufficient quantities of either the modified DNA or individual adduct must be available for immunization and antibody characterization. These limitations may explain why this technique has not been applied in the laboratory or field on a large scale to detect DNA damage in aquatic or terrestrial species. In certain cases, antibodies with broad chemical specificity have been used to detect DNA adducts of a class of carcinogens such as benzo[a]pyrene, benz[a]anthracene and chrysene with application in human health studies (Bartsch et al. 1988; Santella et al. 1987; Santella 1988).

### DNA Strand Breakage

DNA strand breakage is not an uncommon occurrence in a cell. Heat energy causes thousands of abasic sites per cell per day that are rapidly repaired under normal conditions (Elliott and Elliott 1997). This is an example of an insult to DNA that indirectly results in strand breakage (i.e., the initial damage is a loss of a base from the DNA chain, and the repair of this damage results in a temporary gap in the DNA molecule). Ionizing radiation can cause strand breakage directly, whereas other physical agents, such as UV light and chemical agents that are genotoxic, potentiate alterations to the DNA molecule that are candidates for repair (e.g., photoproducts, adducts, etc.) and thus for the occurrence of strand breaks (Shugart et al. 1992; Shugart 1994a).

The reader is referred to the published literature for a detailed explanation of the methodology for performing a particular DNA strand break assay; however, most assays currently in use are based on the general principle that under in vitro conditions, the rate at which single-stranded DNA is released from the duplex DNA at high pH is proportional to the number of strand breaks in the DNA molecule (Rydberg 1975). It should be noted that under these in vitro assay conditions, alkaline labile sites in the DNA molecule (abasic sites for example) will also be detected because they are chemically converted to single strand breaks.

Because the various assays incorporate the same basic principle for determining strand breakage in DNA, their uniqueness has to do with either the manipulation of the DNA preparation before release of the single stranded DNA from the DNA duplex and/or the method for the detection of double-stranded and single-stranded DNA. For example, in both the "alkaline elution" assay of Kohn et al. (1976) and the "DNA precipitation" assay of Olive (1988), denatured single-stranded DNA is released from a physical matrix (fibers constituting the filter of the alkaline elution assay or cellular proteins in the DNA precipitation assay). Because these assays allow for the physical separation of single stranded DNA from double stranded DNA during the denaturation process, quantifying the amount of the two DNA species at the end of the assay is easily accomplished.

In the alkaline unwinding assay (Shugart 1988, 1996a), on the other hand, some degree of DNA purification must be attained before the denaturation step. Furthermore, in this assay both double- and single-stranded DNA species will be present at the end of the denaturation process; therefore, subsequent steps must be incorporated to accommodate this situation. This can be accomplished by either the physical separation of the two species using, for example, hydroxyapatite chromatography (Kanter and Schwartz 1979; Daniel et al. 1985) or by differential extraction with phenol containing 1N NaCl (Morris and Shertzer 1985). A facile procedure that does not require prior separation is based on the difference in fluorescence that occurs when bisbenzimidazole (Hoechst dye No. 33258) binds to the two species of DNA (Kanter and Schwartz 1982; Shugart 1988, 1996a).

Gel electrophoresis represents another analytical technique that may be used to assay for strand breaks in DNA (Theodorakis et al. 1992). Under alkaline conditions, electrophoresis of DNA on agarose gels results in migration within this matrix that is size dependent. Detection is easily accomplished with ethidium bromide staining. With gel electrophoresis it is possible to detect both single strand and double strand breaks (Theodorakis et al. 1994). The latter type of damage to DNA is an extremely important occurrence as it can be a lethal event for most cells. A recent interesting

**TABLE 3.** *Application of DNA strand breakage analysis to genotoxicant exposure*

| Method | Organism | Reference |
|---|---|---|
| Alkaline elution | Marine mussel | Bihari et al. 1990 |
| DNA precipitation | Animal cells | Olive 1988 |
| Alkaline unwinding | Fish | Shugart 1988, 1990b |
| | | Theodorakis et al. 1992 |
| | Rodents | Morris and Shertzer 1985 |
| | Turtles | Meyers-Schone et al. 1993 |
| | Marine mussel | Nacci and Jackim 1989 |
| | Marine seastar | Everaarts et al. 1998 |
| Electrophoresis | | |
| Agarose | Fish | Theodorakis et al. 1994 |
| Single-cell gel | Lymphocytes | Singh et al. 1988 |
| | Animal cells | Olive et al. 1992 |

application of gel electrophoresis assay is the evaluation of DNA strand breakage in individual cells (Singh et al. 1988; Olive et al. 1992). This method is often referred to as the "comet assay" and allows for the observation of subsets of cells with varying levels of DNA strand breakage. A summary of the assay methods currently used to detect DNA strand breakage is found in Table 3. It should be noted that the applications cited are for documentation purposes only.

The interpretation of DNA strand breakage observed in an organism is not a straight-forward matter. As mentioned previously, heat energy within a cell, as well as normal events associated with cellular metabolism, can result in breaks along the backbone of the DNA molecule. In addition, genotoxicants can cause DNA strand breakage. Thus, the problem becomes one of distinguishing between strand breaks that occur under normal conditions from those that result from abnormal situations, such as those that may occur upon environmental exposure to genotoxicants. This distinction can often be accommodated through the selection of a suitable control or reference population for comparison. Ideally, the reference population would not be subject to genotoxic stress. However, this population should experience the same natural environmental stressors as the population under study. Any significant increase in DNA strand breakage in the sampled population over the baseline data observed in the reference population would be indicative of the potential for genotoxic insult to the sampled population. Admittedly, this explanation oversimplifies the interpretative process. For example, many chemicals in the environment may not be genotoxic per se, but can cause necrosis and cell death, i.e., exposure to a nongenotoxic agent initiates a pathological event leading ultimately to the degradation of DNA in the dead cell (Corcoran and Ray 1992). The reader is referred to recently published articles for more information (Shugart 1990b; Shugart et al. 1992; Theodorakis et al. 1992; Shugart 1996b).

In summary, strand breakage in DNA may be used as general biological response to exposure to genotoxicants. That is, the biological response may be indicative of exposure to genotoxicants but does not identify the genotoxicant(s) of concern. Several analytical techniques are available that can detect this type of structural modification to DNA. This approach is readily amenable to field sampling and is currently being applied to genotoxicological problems in the environment (Shugart 1995, 1996b; see Table 3).

## Modified Bases in DNA

### *Minor Nucleoside Content*

In eukaryotic DNA, 5-methylcytosine is generally the only methylated base present in DNA, and its level is enzymatically maintained (Razin and Riggs 1980; Ehrlich and Yang 1981; Holliday 1987). Exposure to certain chemical carcinogens such as benzo[a]pyrene causes hypomethylation of this base in the DNA. Apparently DNA adducts of this genotoxicant affect the specificity of the DNA methylating enzymes (Boehim and Drahovsky 1983; Wilson and Jones 1983; Pfeifer et al. 1984). The base composition of DNA is easily determined by ion-exchange chromatography (Shugart 1990c). As with the detection of DNA strand breakage, this method measures a loss of normal DNA structure (in this case, methylation) but does not identify the chemical responsible.

### *Free Radical Damage to DNA*

The microsomal metabolism of some hydrocarbons produces free radicals, including oxyradicals that are thought to play important roles in chemical carcinogenesis. The biochemical mechanisms whereby fluxes of these radicals are produced in aquatic animals has been reviewed (DiGiulio et al. 1989). Both the superoxide radical and the hydroxyl radical are known to damage DNA directly through strand scission and the oxidation of the bases of DNA, in particular, guanine.

Free radical damage to DNA can result in the hydroxylation of guanine to form 8-hydroxydeoxyguanosine (8-OH-dGuo), which is released from intact DNA when hydrolyzed enzymatically to the deoxynucleoside level. The 8-OH-dGuo is separated from other deoxynucleosides by reverse-phase HPLC, and analysis is with an electro-chemical detector system (Floyd et al. 1986). The level of sensitivity is in the order of one 8-OH-dGuo residue per 105 normal dGuo residues. Free radical damage to the guanine base may also result in the formation of 2,6-diamino-4-hydroxy-5-formamidopyrimidine moiety (FapyGua) in which ring opening of the heterocyclic base occurs. The FapyGua lesion is removed from DNA by acid hydrolysis, trimethyl-silyated, and subjected to gas chromatography-mass spectrometry with single-ion monitoring. Detection of 0.01 nanomoles of this compound in 1 mg of DNA has been reported (Malins et al. 1990).

Results from field studies with English sole (*Parophrys vetulus*) taken from Puget Sound, USA suggested a strong correlation between the levels of chemical contaminants, particularly polycyclic aromatic hydrocarbons (PAHs) and polychlorinated biphenyls (PCBs), in the sediments and the prevalence of liver neoplasms (Malins et al. 1984). Recent studies by Malins and Haimanot (1991) have shown that FapyGua

DNA lesions are also found in the hepatic DNA of English sole from the same environment. The occurrence of this type of DNA damage is indicative of genotoxic insult via a mechanism that produces free radicals.

## Surrogate DNA Adducts

Because evidence of DNA alterations in most tissues, including that of humans, is sometimes difficult to obtain, damage to ancillary molecules may serve as a surrogate (protein adducts for example). Blood proteins, such as hemoglobin and serum albumin, have proven useful in this context (Shugart 1994b).

Because it meets a number of essential requirements, hemoglobin has been proposed as a surrogate for DNA for estimating the in vivo dose of chemicals subsequent to exposure. First, it has reactive nucleophilic sites, and the reaction products with electrophilic agents are stable. Over 60 compounds to date have been shown to yield covalent reaction products with hemoglobin in animal experiments (Wogan and Gorelick 1985; Bartsch et al. 1988). These compounds include representatives of most of the important classes of chemical genotoxicants currently known. No mutagenic or cancer-initiating compound has failed to produce covalent reaction products with hemoglobin. Second, hemoglobin has a well-established lifespan, is readily available in large quantities in humans and animals, and its concentration is not subject to large variation. Third, modification of hemoglobin has been shown to give an indirect measure of the dose to the DNA in cells that are potential targets for genotoxicants (Shugart and Kao 1985).

The detection and measurement of adducts to hemoglobin has been extensively studied in humans exposed to hazardous occupational chemicals for some time, and a considerable scientific literature exists on methodologies and approaches in this area. It is only recently that protein adducts (particularly hemoglobin) have received attention as biomarkers to assess exposure to environmental contamination containing genotoxicants (Shugart et al. 1987; Talmage and Walton, 1991; Blondin and Viau, 1992).

## Other Endpoints and Methods

It was mentioned previously that upon exposure to a genotoxicant, a suite of temporal biological responses may be expressed (Table 1). Most appear shortly after exposure and represent transient structural damage to the DNA molecule. These early occurring responses may or may not persist for a finite period of time. Persistence depends upon factors such as the ability of the organism to repair its damaged DNA and/or the time frame of contact with the genotoxicant (i.e., acute or chronic exposure) as well as dose, etc. However, structural perturbations to the DNA molecule that do persist and do not result in death of the cell in which they occur may potentiate other problems. In particular they may interfere with the fidelity of DNA replication (see discussion below on the flow of genetic stress). If this situation occurs, then the possibility exists for a new suite of biological markers of genotoxicity to be expressed. These endpoints have to do with the appearance of abnormal DNA in the somatic cells of the organism. Abnormal DNA has been associated with clastogenic events such as chromosomal aberrations, sister chromatid exchange, micronuclei formation and aneuploidy, the activation of oncogenes, or the expression of protein dysfunction. Thus, these late occurring responses are more closely linked to whole organism performance and less related to the initial structural damage caused by genotoxicants. Well established techniques and methodologies are available to detect and measure these endpoints (Shugart et al. 1992).

## Significance of Structural Damage to DNA

For a genotoxicant to elicit a response at the molecular or biochemical level, it is necessary that it become bioavailable. There are numerous biological processes that affect the absorption, distribution, and metabolism of a genotoxicant and determine whether or not it will become bioavailable and, if so, the chemical form that interacts with the DNA molecule (Shugart 1996a). Thus, the presence of a specific adduct in the DNA is an indicator of exposure to a genotoxicant and therefore may be used to unequivocally identify the genotoxicant of concern. The occurrence of elevated levels of base modifications (other than adducts) or DNA strand breakage, on the other hand, are general indicators of DNA perturbation arising from exposure to genotoxicants that by themselves possess only inferential value. Structural damage to DNA becomes a powerful diagnostic tool when integrated with other molecular markers or cellular responses to toxicants that occur in the exposed organism (Shugart 1990c, 1996a). Furthermore, characterization of specific DNA adducts may ultimately lead to the identification of a group of genotoxicants of significant environmental concern (Shugart et al. 1992).

It has been assumed that most DNA adducts and other structural alterations to DNA are intrinsically damaging events. This assumption might not be entirely correct because they may be readily repaired without further consequences. Also, adducts on nucleotides in inert regions of DNA may persist without producing adverse effects. Nevertheless, DNA adducts still provide evidence of specific exposure that has passed all of the toxicokinetic barriers. The detection of DNA adducts in organisms collected from their natural habitats provides a means of investigating the qualitative and quantitative relationships between the formation of DNA adducts, subsequent DNA processing, and resulting lesions in target tissues.

It is known that structural damage to DNA that is not properly repaired may result in alterations that become fixed and are eventually transmitted into daughter cells. This flow of "genotoxic stress" (Shugart and Theodorakis 1996) within a somatic cell is generalized in Fig. 2 and the mechanisms involved have been reviewed (Brusick 1980; Thilly and Call 1986); however, several salient points need to be reiterated. Effects to the cell, such as the occurrence of chromosomal aberrations, oncogene activation, and protein dysfunction, are not usually caused by the direct interaction of the genotoxicant with DNA, but rather as the result of a subsequent series of

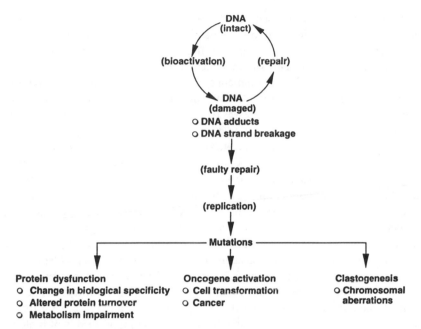

**FIG. 2.** Flow of genetic stress in an organism due to DNA damage (reprinted from Shugart, 1996a, with permission of Lewis Publishers, Boca Raton, FL).

separate and very complex processes for which there is presently only a rudimentary understanding. These processes are affected differently in different species and may depend upon, for example, the type or class of genotoxicant and the reactivity of its metabolite(s), the capacity of the cell to repair DNA damage, and the ability of the cell to recognize and suppress the multiplication of cells with aberrant properties (Thilly and Call 1986). Affected cells often exhibit altered cellular functions indicative of subclinical manifestation of genotoxic disease (Kurelec 1993). Effects expressed in somatic cells can be detrimental to exposed individuals, whereas mutational events in germ cells will affect subsequent generations. Extrapolation of observations made at the level of the somatic cell to predict effects at the level of the germ is difficult. This is due to the inherent difference in sensitivity of these types of cells to genotoxicants (Clive 1987). Furthermore, establishing a causal relationship between a genotoxic agent in the environment and a deleterious effect in subsequent generations of an organism is also highly unlikely because individuals carrying harmful mutations are usually eliminated from the population due to a strong selection against less fit and less well-adapted individuals (Wurgler and Kramers 1992).

## CHALLENGES FOR GENETIC ECOTOXICOLOGY

Genetic ecotoxicology is an approach that applies the principles and techniques of genetic toxicology to assess the potential effects of environmental pollution in the form of genotoxicants on exposed populations. In an attempt to adequately assess the status of the environment and to evaluate the adverse effects of genotoxicants on living systems, ecotoxicology, as a science, must come more and more to rely on a multidisciplinary approach. This situation has arisen for several reasons. First, genotoxicants seldom occur in the environment as a single entity, but rather as mixtures of many different substances whose individual and combined toxicities are modulated by both biological and physical processes (see also Klerks, chapter 6). Second, living systems occur as a continuum from simple organisms to multiple species that interact not only among themselves in a complicated manner, but also with their physical environment. Third, an understanding of complex toxicological phenomena requires that investigations be conducted at different levels of biological organization (i.e., molecular, cellular, individual, population, community, ecosystem), in many environments (aquatic, terrestrial, polar, and so forth), and where genotoxicants, both natural and anthropogenic, are present. Thus, success in genetic ecotoxicological investigations dictates the integration of the unique knowledge of other scientific disciplines and the implementation of their sophisticated approaches and techniques for problem solving.

Genetic toxicologic investigations are important for the documentation of exposure to genotoxicants; however, they often fail to provide the information necessary to establish the outcome for the environment. There are many reasons for this, but one very important one is that the responses observed at higher levels of biological organization are latent and so far removed from the initial events of exposure that causality is difficult to establish. A way to approach this problem is to ask mechanistic questions about how an organism relates to its environment. Ecosystems result from the dynamic interaction of living and inert matter in which the living material acclimates and adapts to environmental change. Many of the processes occurring within living organisms, including responses to genotoxicants, have a genetic basis (Shugart and Theodorakis 1996). Therefore, understanding changes at the genetic level (DNA) should help define the more complex changes at the ecosystem level and in all probability will constitute the major thrust in the field of genetic ecotoxicology (Anderson et al. 1994).

## REFERENCES

Anderson, S., W. Sadinski, L. Shugart, P. Brussard, M. Depledge, T. Ford, J. Hose, J. Stegeman, W. Suk, I. Wirgin, and G. Wogan. 1994. Genetic and molecular ecotoxicology: A research framework. *Environmental Health Perspectives* 102:3–8.

Bartsch, H., K. Hemminki, and K. O'Neill. 1988. *Methods for detecting DNA damaging agents in humans: Application in cancer epidemiology and prevention.* IARC Scientific Publication No. 89. Lyon, France: IRAC.

Bickham, J. W., and M. J. Smolen. 1994. Somatic and heritable effects of environmental genotoxins and the emergence of evolutionary toxicology. *Environmental Health Perspectives* 12:25–28.

Bihari, N., R. Batel, and R. K. Zahh. 1990. DNA damage determination by alkaline elution technique in the hemolymph of mussel *Mytilus galloprovincialis* treated with benzo[a]pyrene and 4-nitroquinolin-N-oxide. *Aquatic Toxicology* 18:13–18.

Blondin, O., and C. Viau. 1992. Benzo[a]pyrene protein adducts in wild woodchucks used as biological sentinels of environmental PAH contamination. *Archives Environmental Contamination and Toxicology* 3:310–315.

Boehim, T. L., and D. Drahovsky. 1983. Alteration of enzymatic methylation of DNA cystosines by chemical carcinogens: A mechanism involved in the initiation of carcinogenesis. *Journal of the National Cancer Institute* 71:429–433.

Brusick, D. 1980. *Principles of genetic toxicology.* New York: Plenum Press.

Clive, D. 1987. Genetic toxicology: From theory to practice. *Clinical Research and Drug Development* 1:11–41.

Corcoran, G. B., and S. D. Ray. 1992. The role of the nucleus and other compartments in toxic cell death produced by alkylating hepatotoxicants. *Toxicology and Applied Pharmacology* 113:167–171.

Daniel, F. B., D. L. Haas, and S. M. Pyle. 1985. Quantitation of chemically induced DNA strand breaks in human cells via an alkaline unwinding assay. *Analytical Biochemistry* 144:390–402.

Das, R. K., and N. K. Nanda. 1986. Induction of micronuclei in peripheral erythrocytes of fish *Heteropneustes fossilis* by mitomycin C and paper mill effluent. *Mutation Research* 175:67–71.

DiGiulio, R. T., P. C. Washburn, R. J. Wenning, G. W. Winston, and C. C. Jewell. 1989. Biochemical responses in aquatic animals: A review of determinants of oxidative stress. *Environmental Toxicology and Chemistry* 8:1103–1123.

Dixon, D. R. 1982. Aneuploidy in mussel embryos (*Mytilus edulis* L.) originates from a polluted dock. *Marine Biology Letters* 3:155–161.

Dunn, B. P., J. Black, and A. Maccubbin. 1987. [32]P-Postlabeling analysis of aromatic DNA adducts in fish from polluted areas. *Cancer Research* 47:6543–6548.

Dunn, B. P., and R. H. C. San. 1988. HPLC enrichment of hydrophobic DNA-carcinogen adducts for enhanced sensitivity of [32]P-postlabeling analysis. *Carcinogenesis* 9:1055–1060.

Ehrlich, M., and R. Y.-H. Yang. 1981. 5-methylcytosine in eukaryotic DNA. *Science* 212:1350–1353.

El-Adlouni, C., J. Tremblay, P. Walsh, J. Lageux, J. Bureau, D. Laliberte, G. Keith, D. Nadeau, and G. G. Poirier. 1995. Comparative study of DNA adduct levels in white sucker fish (*Catostomus commersoni*) from the basin of the St. Lawrence river (Canada). *Molecular Cell Biochemistry* 148:133–138.

Elliott, W. H., and D. C. Elliott. 1997. *Biochemistry and molecular biology.* New York: Oxford University Press.

Everaarts, J. M., P. J. den Besten, M. Th. J. Hillebrand, R. S. Halbrook, and L. R. Shugart. 1998. DNA strand breaks, cytochrome P450 dependent monooxygenase system activity and levels of chlorinated biphenyl congeners in the pyloric caeca of the seastar (*Asterias rubens*) from the North Sea. *Ecotoxicology* 7: 69–79.

Floyd, R. A., J. J. Watson, P. K. Wong, D. H. Altmiller, and R. C. Rickard. 1986. Hydroxyl free radical adduct of deoxyguanosine: Sensitive detection and mechanisms of formation. *Free Radical Research Communications* 1:163–172.

French, B. L., W. L. Reichert, T. Hom, H. R. Nishimoto, and J. E. Stein. 1996. Accumulation and dose-response of hepatic DNA adducts in English sole (*Pleuronectes vetulus*) exposed to a gradient of contaminated sediments. *Aquatic Toxicology* 36:1–16.

Gupta, R. C. 1985. Enhanced sensitivity of [32]P-postlabeling analysis of aromatic carcinogen-DNA adducts. *Cancer Research* 45:5656–5662.

Gupta, R. C., and K. Randerath. 1988. Analysis of DNA adducts by [32]P-labeling and thin layer chromatography. In *DNA repair*, vol. 3, eds. E. C. Friedberg and P. C. Hanawalt, 399–418. New York: Marcel Dekker, Inc.

Gupta, R. C., M. C. Reddy, and K. Randerath. 1982. [32]P-labeling analysis of nonradioactive aromatic carcinogen-DNA adducts. *Carcinogenesis* 3:1081–1092.

Halbrook, R. S., R. L. Kirkpartick, D. R. Bevan, and B. P. Dunn. 1992. DNA adducts detected in muskrats by [32]P-Postlabeling analysis. *Environmental Toxicology and Chemistry* 11:1605–1611.

Harvey, R. C. 1982. Polycyclic hydrocarbons and cancer. *American Scientist* 70:386–393.

Holliday, R. 1987. The inheritance of epigenetic defects. *Science* 238:163–166.

Hose, J. E., J. N. Cross, S. C. Smith, and D. Diehl. 1987. Elevated circulating erythrocyte micronuclei in fishes from contaminated sites off Southern California. *Marine Environmental Research* 22: 167–176.

Kanter, P. M., and H. A. S. Schwartz. 1979. A hydoxylapatite batch assay for quantitation of cellular DNA damage. *Analytical Biochemistry* 97:77–84.

Kanter, P. M., and H. A. S. Schwartz. 1982. A fluorescence enhancement assay for cellular DNA damage. *Molecular Pharmacology* 22:145–151.

Kohn, K. W., L. C. Erickson, A. G. Ewig, and C. A. Friedman. 1976. Fractionation of DNA from mammalian cells by alkaline elution. *Biochemistry* 15:4629–4637.

Kurelec, B. 1993. The genotoxic disease syndrome. *Marine Environmental Research* 35:341–348.

Kurelec, B., A. Garg, S. Krca, M. Chacko, and R. C. Gupta. 1989. Natural environment surpasses polluted environment in inducing DNA damage in fish. *Carcinogenesis* 7:1337–1339.

Kurelec, B., A. Garg, S. Krca, and R. C. Gupta. 1990. DNA adducts in marine mussels *Mytilus galloprovincialis* living in polluted and unpolluted environments. In *Biomarkers of environmental contamination*, eds. J. F. McCarthy and L. R. Shugart, 217–227. Boca Raton, FL: Lewis Publishers Inc.

Kurelec, B., A. Garg, S. Krca, S. Britvic, D. Lucic, and R. C. Gupta. 1992. DNA adducts in carp exposed to artificial diesel-2 oil slicks. *European Journal of Pharmacology* 1:51–56.

Maddock, M. B., H. Northrup, and T. J. Ellingham. 1986. Induction of sister-chromatid exchanges and chromosomal aberrations in hematopoietic tissue of marine fish following in vivo exposure to genotoxic carcinogens. *Mutation Research* 172:165–175.

Malins, D. C., and R. Haimanot. 1991. The etiology of cancer: Hydroxyl radical-induced DNA lesions in histologically normal livers of fish from a population with liver tumors. *Aquatic Toxicology* 20: 123–130.

Malins, D. C., B. B. McCain, D. W. Brown, S. L. Chan, M. S. Myers, J. T. Landahl, P. G. Prohaska, A. J. Friedman, L. D. Rhodes, D. G. Burrows, W. D. Gronlund, and H. O. Hodgins. 1984. Chemical pollutants in sediments and diseases in bottom-dwelling fish in Puget Sound, Washington. *Environmental Science and Technology*. 18:705–713.

Malins, D. C., G. K. Ostrander, R. Haimanot, and P. Williams. 1990. A novel DNA lesion in neoplastic livers of feral fish: 2,6,-diamino-4-hydroxy-5-formamidopyrimidine. *Carcinogenesis* 11:1045–1047.

Marnett, L. J. 1993. Frontiers in molecular toxicology. *Chemical Research in Toxicology* 6:739–740.

Martineau, D., A. Legace, P. Beland, R. Higgins, D. Armstrong, and L. R. Shugart. 1988. Pathology of stranded beluga whales *Delphinapterus leucas* from the St. Lawrence estuary, Quebec, Canada. *Journal of Comparative Pathology* 9:287–300.

McCarthy, J. F., H. Gardner, M. J. Wolfe, and L. R. Shugart. 1991. DNA alterations and enzyme activities in Japanese medaka (*Oryzias latipes*) exposed to diethylnitrosamine. *Neuroscience and Biobehaviorial Reviews* 15:99–102.

McMahon, G., L. J. Huber, M. J. Moore, J. J. Stegeman, and G. N. Wogan. 1990. Mutations in c-K-ras oncogenes in diseased livers of winter flounder from Boston Harbor. *Proceedings of the National Academy of Science USA* 87:841–845.

Meyers-Schone, L., L. R. Shugart, J. J. Beauchamp, and B. T. Walton. 1993. Comparison of two freshwater turtle species as monitors of chemical contamination: DNA damage and residue analysis. *Environmental Toxicology and Chemistry* 12:1487–1496.

Morris, S. R., and H. G. Shertzer. 1985. Rapid analysis of DNA strand breaks in soft tissues. *Environmental Mutagenesis* 7:871–880.

Nacci, D., and E. Jackim. 1989. Using the DNA alkaline unwinding assay to detect DNA damage in laboratory and environmentally exposed cells and tissues. *Marine Environmental Research* 28:333–336.

Olive, P. L. 1988. DNA precipitation assay: A rapid and simple method for detecting DNA damage in mammalian cells. *Environmental Molecular Mutagenesis* 11:487–495.

Olive, P. L., D. Wlodek, R. E. Durand, and J. P. Banath. 1992. Factors influencing DNA migration from individual cells subjected to gel electrophoresis. *Experimental Cell Research* 198:259–266.

Pfeifer, G. P., D. Grungerger, and D. Drahovsky. 1984. Impaired enzymatic methylation of BPDE- modified DNA. *Carcinogenesis* 5:931–936.

Phillips, D., and P. Sims. 1979. PAH metabolites: Their reaction with nucleic acids. In *Chemical carcinogens and DNA, vol. 2*, ed. P. L. Grover, 9–57. Boca Raton, FL: CRC Press.

Qu, S.-X., C.-L. Bai, and N. H. Stacey. 1997. Determination of bulky DNA adducts in biomonitoring of carcinogenic chemical exposures: Features and comparison of current techniques. *Biomarkers* 2: 3–6.

Rahn, R., S. Chang, J. M. Holland, and L. R. Shugart. 1982. A fluorometric-HPLC assay for quantitating the binding of benzo[a]pyrene metabolites to DNA. *Biochemical and Biophysical Research Communications* 109:262–268.

Randerath, K., M. V. Reddy, and R. C. Gupta. 1981. 32P-Labeling test for DNA damage. *Proceedings of the National Academy of Science USA* 78:6126–6129.

Randerath, K., M. C. Reddy, and R. M. Disher. 1986. Age- and tissue-related DNA modifications in untreated rats: Detection by $^{32}$P-postlabeling assay and possible significance for spontaneous tumor induction and aging. *Carcinogenesis* 7:1615–1619.

Ray, S., B. P. Dunn, J. F. Payne, L. Fancey, R. Helbig, and P. Beland. 1991. Aromatic DNA-carcinogen adducts in beluga whales from the Canadian Arctic and gulf of St. Lawrence. *Marine Pollution Bulletin* 22:392–395.

Razin, A., and A. D. Riggs. 1980. DNA methylation and gene function. *Science* 210:604–607.

Reddy, M. V., and K. Randerath. 1986. Nuclease P1-mediated enhancement of sensitivity of [32]P-postlabeling test for structurally diverse DNA adducts. *Carcinogenesis* 7:1543–1551.

Rydberg, B. 1975. The rate of strand separation in alkali of DNA of irradiated mammalian cells. *Radiation Research* 61:274–285.

Sandhu, S. S., and W. R. Lower. 1989. In situ assessment of genotoxic hazards of environmental pollution. *Toxicology and Industrial Health* 5:73–83.

Santella, R. M. 1988. Application of new techniques for the detection of carcinogen adducts to human population monitoring. *Mutation Research* 205:271–282.

Santella, R. M., F. Gasparo, and L. Hsieh. 1987. Quantitation of carcinogen-DNA adducts with monoclonal antibodies. *Progress in Experimental Tumor Research* 31:63–75.

Shugart, L. R. 1988. Quantitation of chemically induced damage to DNA of aquatic organisms by alkaline unwinding assay. *Aquatic Toxicology* 13:43–52.

Shugart, L. R. 1990a. Biological monitoring: Testing for genotoxicity. In *Biological markers of environmental contaminants*, eds. J. F. McCarthy and L. R. Shugart, 205–216. Boca Raton, FL: Lewis Publishers.

Shugart, L. R. 1990b. DNA damage as an indicator of pollutant-induced genotoxicity. In *13th symposium on aquatic toxicology and risk assessment: Sublethal indicators of toxic stress*, eds., W. G. Landis and W. H. van der Schalie, 348–355. Philadelphia: ASTM.

Shugart, L. R. 1990c. 5-methyl deoxycytidine content of DNA from Bluegill Sunfish *Lepomis macrochirus* exposed to benzo[a]pyrene. *Environmental Toxicology and Chemistry* 9:205–208.

Shugart, L. R. 1994a. Genotoxic responses in blood. In *Nondestructive biomarkers in vertebrates*, eds. M. C. Fossi and C. Leonzio, 131–145. Boca Raton, FL: Lewis Publishers.

Shugart, L. R. 1994b. Hemoglobin adducts. In *Nondestructive biomarkers in vertebrates*, eds. M. C. Fossi and C. Leonzio, 159–168. Boca Raton, FL: Lewis Publishers.

Shugart, L. R. 1995. Environmental genotoxicology. In *Fudamentals of aquatic toxicology, second edition*, ed. G. M. Rand, 405–420. Washington, DC: Taylor & Francis Publishers.

Shugart, L. R. 1996a. Molecular markers to toxic agents. In *Ecotoxicology a hierarchical treatment*, eds. M. C. Newman and C. H. Jagoe, 133–161. Boca Raton, FL: Lewis Publishers.

Shugart, L. R. 1996b. Application of the alkaline unwinding assay to detect DNA strand breaks in aquatic species. In *Techniques in aquatic toxicology*, ed. G. K. Ostrander, 205–218. Boca Raton, FL: CRC Press.

Shugart, L. R., and J. Kao. 1985. Examination of adduct formation in vivo in the mouse between benzo[a]pyrene and DNA of skin and hemoglobin of red blood cells. *Environmental Health Perspectives* 62:223–226.

Shugart, L. R., and C. Theodorakis. 1996. Genetic ecotoxicology: The genotypic diversity approach. *Comparative Biochemistry and Physiology* 113C:273–276.

Shugart, L. R., J. M. Holland, and R. Rahn. 1983. Dosimetry of PAH carcinogenesis: Covalent binding of benzo[a]pyrene to mouse epidermal DNA. *Carcinogenesis* 4:195–198.

Shugart, L. R., J. M. McCarthy, B. D. Jimenez, and J. Daniel. 1987. Analysis of adduct formation in the Bluegill Sunfish *Lepomis macrochirus* between benzo[a]pyrene and DNA of the liver and hemoglobin of the erythrocyte. *Aquatic Toxicology* 9:319–324.

Shugart, L. R., J. Bickham, G. Jackim, G. McMahon, W. Ridley, J. Stein, and S. Steiner. 1992. DNA alterations, In *Biomarkers: Biochemical, physiological and histological markers of anthropogenic stress*, eds. R. J. Huggett, R. A. Kimerle, P. M. Mehrle, and H. L. Bergman, 125–153. Boca Raton, FL: Lewis Publishers.

Singh, N. P., M. T. McCoy, R. R. Tice, and E. L. Schneider. 1988. A simple technique for quantitation of low levels of DNA damage in individual cells. *Experimental Cell Research* 175:184–191.

Talmage, S. S., and B. T. Walton. 1991. Small mammals as monitors of environmental contaminants. *Reviews in Environmental Toxicology* 119:47–145.

Theodorakis, C. W., S. J. D'Surney, J. W. Bickham, T. B. Lyne, B. P. Bradley, W. E. Hawkins, W. L. Farkas, J. F. McCarthy, and L. R. Shugart. 1992. Sequential expression of biomarkers in bluegill sunfish exposed to contaminated sediment. *Ecotoxicology* 1:45–73.

Theodorakis, C. W., S. J. D'Surney, and L. R. Shugart. 1994. Detection of genotoxic insult as DNA strand breaks in fish blood cells by agarose gel electrophoresis. *Environmental Toxicology and Chemistry* 7:1023–1031.

Thilly, W. G., and K. M. Call. 1986. Genetic toxicology. In *Third edition of Casarett and Doull's toxicology*, eds. D. D. Klaass, M. O. Amdur, and J. Doull, 174–194. New York: Macmillan Publishing Co.

van Schooten, F. J., L. M. Maas, E. J. C. Moonen, and R. van der Oost. 1995. DNA adduct dosimetry in biological indicator species living on PAH-contaminated soils and sediments. *Ecotoxicology and Environmental Safety* 30:171–179.

Varanasi, U., W. L. Reichert, and J. E. Stein. 1989. [32]P-postlabeling analysis of DNA adducts in liver of wild English Sole *Paraophrys vetulus* and winter flounder *Pseudopleuronectes americanus. Cancer Research* 49:1171–1177.

Walsh, P., C. El-Adlouni, M. J. Mukhopadhyay, G. Viel, D. Nadeau, and G. G. Poirier. 1995. [32]P-Postlabeling determination of DNA adducts in earthworm *Lumbricus terrestris* exposed to PAH-contaminated soils. *Bulletin of Environmental Contamination and Toxicology* 54:654–661.

Weinstein, I. B. 1978. Current concepts on mechanisms of chemical carcinogenesis. *Bulletin of the New York Academy of Medicine* 54:336–383.

West, W. R., P. A. Smith, G. M. Booth, and M. L. Lee. 1988. Isolation and detection of genotoxic components in a Black River sediment. *Environmental Science and Technology* 22:224–228.

Wilson, V. L., and P. A. Jones. 1983. Inhibition of DNA methylation by chemical carcinogens in vitro. *Cell* 32:229–238.

Wogan, G. N., and N. J. Gorelick. 1985. Chemical and biochemical dosimetry to exposure to genotoxic chemicals. *Environmental Health Perspectives* 62:5–18.

Wurgler, F. E., and P. G. N. Kramers. 1992. Environmental effects of genotoxins (eco-genotoxicology). *Mutagenesis* 7:321–327.

Yung-Jin, C., C. Mathews, K. Mangold, K. Marien, J. Hendricks, and G. Bailey. 1991. Analysis of ras gene mutations in rainbow trout liver tumors initiated by aflatoxin B1. *Molecular Carcinogenesis* 4:112–119.

*Genetics and Ecotoxicology*
Edited by V. E. Forbes
Copyright © 1999 Taylor & Francis

# 9

# Toxicity and Sexual Reproduction in Rotifers: Reduced Resting Egg Production and Heterozygosity Loss

Terry W. Snell, Manuel Serra, and Maria Jose Carmona

**Abstract.** The effect of toxicants on the genetic structure of populations is not well understood. Toxicant exposures often cause strong directional selection and concomitant increased genetic drift at nonselected loci that could alter a population's probability of extinction. Current toxicity tests with laboratory or natural populations do not consider such effects. The main long-term impacts of toxicity on sexual reproduction in rotifer zooplankters are reduced resting egg production and loss of heterozygosity. Using computer simulation of rotifer population dynamics, we explored how toxicity alters genetic structure through reductions in population growth rate, mictic ratio, and fertilization rate, and how these changes affect loss of heterozygosity. The relationship between resting egg production and toxicant-caused reductions in population growth rate ($r$) is a negative exponential, so small reductions in $r$ lead to large reductions in resting egg production. Reductions in the mictic ratio, the proportion of daughters produced by a female that are mictic, also caused exponential decreases in resting egg production. The rate of heterozygosity loss is directly related to effective population size. In rotifers, effective population size ($N_e$) is determined by the volume from which a mictic female obtains her mate, because this determines the number of randomly-mating males and females participating in resting egg production. For mating volumes greater than 100 liters, $N_e$ is so large that heterozygosity loss is not accelerated by reductions in $r$, mictic ratio, or any other parameter investigated. Only when mating volume is less than 10 liters ($N_e$=18) do toxicant-induced reductions in these parameters accelerate heterozygosity loss. However, under all conditions tested, heterozygosity loss due to toxicant-caused reductions in population growth rate are small. What is currently being measured by standard toxicity tests may not be very relevant for predicting the long-term survival of zooplankton populations because resting egg production is not considered. Yet resting eggs are essential for the long-term persistence of rotifer populations in most environments. By emphasizing the development of convenient, rapid toxicity tests, ecotoxicologists may have overlooked variables that have the greatest impact on the evolutionary fate of zooplankton populations.

**Keywords.** Computer simulation; ecotoxicity tests; effective population size; mictic ratio; population dynamics.

## INTRODUCTION

The effects of toxicants on growth, survival, and reproduction have been intensely studied in many species. Less well known is how toxicants modify a population's genetic structure through selection and genetic drift. One well-documented effect on genetic structure is change in genotypic composition by selection for toxicant resistance. Selection for tolerant genotypes has been reported in several groups, including fish (Guttman 1994), insects (Maroni et al. 1987; Roush and McKenzie 1987; Postma and Groenendijk chapter 5), oligochaetes (Klerks and Levinton 1989), polychaetes (Grant et al. 1989), and several species of plants (Shaw chapter 2). Furthermore, substantial genetic variation for responses to selection for pollution resistance seems to be the norm in most populations (Hoffmann and Parsons 1991), including zooplankton (Baird and Barata chapter 11). The effects of strong directional selection for toxicant resistance on a population are not entirely understood, but probably contribute to the loss of genetic diversity (Forbes 1996). Zooplankton with cyclically parthenogenetic life cycles like rotifers and cladocerans may be especially vulnerable to loss of genetic variability through selection. Clonal selection during parthenogenetic reproduction is particularly efficient since it operates on the genome as a whole and not just on the additive component of genetic variance (Wright 1977; Bell 1982). Rapid changes in the genotypic composition of rotifer populations have been observed in response to cyanobacterial toxicity (Snell 1980).

Genetic variability is essential for adaptation to environmental change (Fisher 1930). Loss of genetic variability from a population, therefore, could be an important change in its genetic structure, potentially reducing chances for long-term survival. A challenge for those interested in genetics and ecotoxicology is to assess how toxicity modifies selection and genetic drift and determines the direction and rate of gene frequency change. A key characteristic of a population determining the impact of genetic drift is its effective population size (Crow and Kimura 1970). Any effects of toxicants that reduce effective population size would increase the level of inbreeding. Strong inbreeding has been shown to have deleterious effects on the fitness of several *Daphnia* species (Innes 1989; DeMeester 1993; Lynch and Deng 1994). The long-term consequences of strong directional selection and increased genetic drift in populations are not well understood, but could alter the probability of extinction. Current tests to assess the impact of toxicity on laboratory or natural populations do not consider such effects (Depledge 1994, 1996).

In rotifer and cladoceran zooplankters, sexual reproduction occurs intermittently in an otherwise parthenogenetic life cycle (Fig. 1) and is triggered by specific environmental signals (Snell and Boyer 1988; Kleiven et al. 1992). Toxicants can modify genetic structure through differential effects on the sexual and asexual phases of the life cycle. For example, toxicity can alter dispersal by slowing male or female swimming speed or changing its direction. Reduction in dispersal distances can alter the volume from which mates are derived and therefore effective population size.

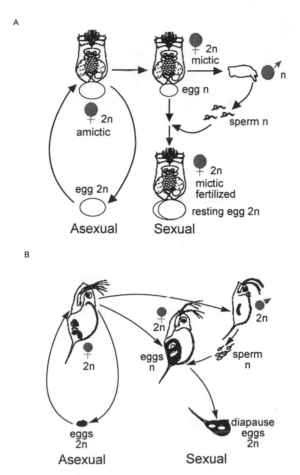

**FIG. 1.** The cyclical parthenogenetic life cycle: (*a*) rotifers (from Snell and Carmona 1995); and (*b*) *Daphnia* (modified from Mort 1991).

Similarly, if toxicant stress reduces the length of the male or female fertile period, mates will come from a smaller volume and effective population size will again be reduced. Toxicant-caused reduction of reproductive rate can reduce population density during sexual reproduction, reducing effective population size. One way that selection could modify genetic structure is by operating differently in males and females. This is likely in rotifers where males are haploid and females are diploid (Birky and Gilbert 1971; Wallace and Snell 1991). Normally recessive, deleterious alleles are not masked by dominance in male rotifers and therefore can be effectively eliminated by selection. Lastly, toxicants could differentially affect the sexual phase of the cyclical parthenogenetic life cycle, changing the frequency or amount of sexual reproduction. The frequency of recombination has a marked effect on the genotypic composition

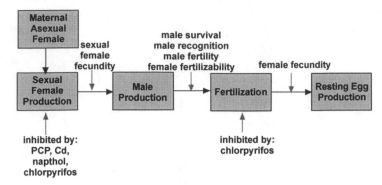

**FIG. 2.** Processes in rotifer sexual reproduction that are inhibited by toxicants.

of a rotifer population (King 1993). If asexual reproduction becomes more dominant, rates of adaptation to environmental change could be modified.

The work of Snell and Carmona (1995) showed that toxicants can have a differential effect on asexual and sexual reproduction in rotifers. Of the four toxicants investigated (pentachlorophenol, cadmium, naphthol, and chlorpyrifos) all reduced the production of sexual females (Fig. 2). Sexual female production can be reduced in at least two ways. One is to increase the mictic threshold, the density of rotifers required to initiate sexual reproduction. A second is to decrease the mictic ratio, the proportion of a female's daughters that reproduce sexually. In addition to inhibiting sexual female production, sexual reproduction can be diminished by inhibition of fertilization. Chlorpyrifos was observed to inhibit fertilization, although the mechanism by which this occurs is unknown. There are clearly many other potential sites for toxicants to differentially inhibit sexual reproduction. More work is necessary to describe toxicant effects on these processes, which are critical for sexual reproduction in rotifers.

The main long-term impacts of toxicity on sexual reproduction in rotifers are likely to be reduced resting egg production and loss of heterozygosity. Investigations of reduction of the resting egg pool and heterozygosity loss are very difficult experimentally. Consequently, we have taken a simulation approach using a model derived from time series analysis of the dynamics of natural rotifer populations (Snell and Serra 1998). A consensus model was derived from the examination of nine rotifer species that proved reliable in predicting the complex dynamics of these natural populations. This model incorporated dynamics from only asexual reproduction, so in this work we added sub-models to describe the dynamics of resting egg production and heterozygosity. Due to the importance of both resting egg production and heterozygosity to the long-term survival of rotifer populations, we have focused our analyses on these variables.

Our objectives in this paper are to explore how toxicity might alter the genetic structure of rotifer populations, to examine how toxicant induced reductions in population growth rate, mictic ratio, and fertilization rate, and increased mictic thresholds affect resting egg production, to investigate how these same changes affect loss of heterozygosity, and to determine how altered genetic structure might influence the evolutionary fate of rotifer populations.

## METHODS

According to Snell and Serra (1998), we simulated the dynamics of rotifer populations over one growing season using the following function for the per capita growth rate $r(t)=$Ln$(N(t)-Ln(t-T))$:

$$r(t) = a_0 + a_1/N(t - T) + a_2/N(t - T)^2 + e(t) \tag{1}$$

where $N(t-T)$ is the population density at time $t-T$, $T$ is the sampling frequency, $a_0$, $a_1$, and $a_2$ are constants, and $e(t)$ is a random variable with distribution $N(0, \sigma^2)$. This function was derived from a commercially available program from Applied Biomathematics (Setauket, New York) called Ramas/time, which gave a satisfactory description for time series of rotifer density recorded in natural populations. One rotifer growing season is defined as hatch from resting eggs in the spring through population die off in the fall. For our simulation, this period was defined as 240 days. All of the within-year dynamics are due to parthenogenetic population growth. When resting eggs are produced they fall to the bottom and remain dormant until the following spring. The parameters in this model were estimated by Snell and Serra from a time series for the population density of the rotifer *Asplanchna girodi* in Golf Course Pond near Tampa, Florida in 1977 (King and Snell, 1980). In that experimental study, female density was recorded every other day for 53 samplings (106 days); thus, according to this sampling frequency, $T=2$ days. Other parameters in the model as estimated by Ramas/time are $a_0=-0.305$, $a_1=0.081$, $a_2=-0.001$, and $\sigma^2=0.786$ (parameter units: $2d^{-1}$). In the simulations, initial density was one female/liter and increased by parthenogenetic reproduction. In the experimental dynamics, changes in the per capita growth rate were implemented by modifying $a_0$, the density-independent term in our model. A population was considered extinct when its density fell below 0.01 females/liter, and the simulation was terminated. Likewise, the simulation was terminated when a population persisted for 240 days, the maximum duration of a growing season.

In order to estimate resting egg production, a relationship between population density and the dynamics of sexual reproduction is needed. We modeled this relationship as follows: we assumed that sexual reproduction (mixis) occurred when a threshold population density was achieved. According to the estimations reported in the literature, an appropriate value for this parameter in *Asplanchna* is five females/liter (Table 1). This was the mictic threshold in the absence of toxicity. If mixis was induced, a proportion of the females were assumed to be mictic females (mictic ratio), i.e., females producing meiotic eggs that if fertilized become resting eggs. Unfertilized eggs develop into haploid males (Fig. 1). Mictic ratio in the absence of toxicity was assumed to be 0.25 (Table 1). The dynamics of sexual reproduction was modeled for

**TABLE 1.** *Experimental values for model parameters*

| Parameter | Species/Strain | Source | Value mean (range) | Reference |
|---|---|---|---|---|
| Mictic threshold (model = 5/L) | *A. priodonta* | nature | (5–6 ind/l) | Ruttner 1930 |
| | *A. priodonta* | nature | (10–100 ind/l) | Carlin 1943 |
| | *A. priodonta* | model | (0.51–1.42 ind/l) | Gerritsen 1980 |
| | *B. calyciflorus* | model | (1.24–7.46 ind/l) | Gerritsen 1980 |
| | *B. angularis* | model | (1.24–7.46 ind/l) | Gerritsen 1980 |
| | *A. girodi* | nature | 2.3 ind/l | King & Snell 1980 |
| | *B. plicatilis* | model | 11.7 mictic fem/l (1–25) | Snell & Garman 1986 |
| | *B. plicatilis* (RUS) | lab | 0.147 females/l | Snell & Boyer 1988 |
| | *B. plicatilis* | nature | 30 females/l | Carmona et al. 1995 |
| | *B. rotundiformis* (SS) | nature | 90 females/l | Carmona et al. 1995 |
| Mictic ratio (model = 0.25) | *A. sieboldi* (10C6) | lab | (0.11–0.69)% mictic fem | Gilbert 1975 |
| | *A. brightwelli* ($5B4S_81$) | lab | (0.21–0.35) | Gilbert 1975 |
| | *A. girodi* (5A1) | lab | (0.08–0.62) | Gilbert & Litton 1978 |
| | *A. girodi* (7C25) | lab | (0.17–0.26) | Gilbert & Litton 1978 |
| | *B. calyciflorus* (GC) | lab | 0.54 (0.18–0.89) | Pourriot & Rougier 1977 |
| | *B. plicatilis* (GS74) | lab | 0.32 (0.07–0.5) | Pourriot & Rougier 1979 |
| | *B. plicatilis* (RKE) | lab | 0.03 | Ruttner-Kolisko 1985 |
| | *B. plicatilis* (RKL) | lab | 0.01 | Ruttner-Kolisko 1985 |
| | *B. plicatilis* (8105A) | lab | (0.015–0.05) | Hagiwara & Hino 1989 |
| | *B. plicatilis* (CU7) | nature | 0.11 | Carmona et al. 1994 |
| | *B. plicatilis* | nature | (0.05–0.23) | Carmona et al. 1995 |
| | *B. rotundiformis* (SS) | lab | (0.05–0.2) | Carmona et al. 1995 |
| | *B. plicatilis* (L1) | lab | 0.32 | Gomez et al. 1997 |
| | *B. rotundiformis* (SM2) | lab | 0.42 | Gomez et al. 1997 |
| | *B. rotundiformis* (SS2) | lab | 0.12 | Gomez et al. 1997 |
| Encounter rate (model = 0.015) | *B. plicatilis* | model | 0.015 encounters/ min | Snell & Garman 1986 |
| Resting egg production (model = 3 RE/ female) | *Euchlanis dilatata* | lab | (3.1–6.3) RE/ female | King 1970 |
| | *Brachionus calyciflorus* | lab | (2–3) | Ruttner-Kolisko 1974 |
| | *B. plicatilis* (RUS) | lab | 2.5 | Snell & Childress 1987 |
| | *B. plicatilis* (L1) | lab | (1–6) | Gomez 1996 |
| | *B. rotundiformis* (SM2) | lab | 1 | Gomez 1996 |
| | *B. rotundifromis* (SS2) | lab | (1–6) | Gomez 1996 |
| Unfertilized mictic female fecundity (model = 3 males/ day) | *Euchlanis dilatata* | lab | (1.9–5.0) males/day | King 1970 |
| | *Brachionus plicatilis* (CU) | lab | 2.2 | Serra 1987 |
| | *B. rotundiformis* (SPO) | lab | 2.8 | Serra 1987 |
| | *Asplanchna brightwelli* | lab | 2.5 | Snell, unpublished |

one-day intervals. A mictic female at time $t$ was assumed to have a probability of being fertilized given by a Poisson distribution (Snell and Garman 1986):

$$1 - \exp(-epM(t)) \tag{2}$$

where $e$ is the mating encounter rate ($0.015Ld^{-1}$, from Snell and Garman 1986), $p$ is the time for encounters in days, and $M(t)$ is the density of males per liter at time $t$. If fertilized, a mictic female produces an average of three resting eggs (Table 1), which remain unhatched in the sediment until a later growing season (usually the next year). $M(t)$ was estimated from the density of unfertilized mictic females at time $t-1$, the probability of remaining unfertilized being $1-$ the probability of being fertilized (Eq. 2). We used an unfertilized mictic female fecundity of three males per female per day (Table 1). The effects of toxicants in reducing the fertilization rate were simulated by multiplying Eq. 2 by a factor corresponding to the probability of being fertilized. For example, if toxicity reduced fertilization rate 10%, the probability of fertilization dropped from 1.0 to 0.9.

We computed the heterozygosity loss due to gamete sampling that is associated with resting egg production. We used

$$H(t)/H = (1 - 1/[2N_e(t)]) \tag{3}$$

(Hedrick 1984), where $H$ is the heterozygosity in the parthenogenetic population, assumed constant for the growth period (resting egg hatching was assumed absent during the growth cycle), $H(t)$ is the heterozygosity in the resting eggs produced at time $t$ in a single growing season, and $N_e(t)$ is the effective population size of the population producing the resting eggs at time $t$. Given that the resting eggs are carrying genes that are a sample from genes carried by unfertilized mictic females (those producing males), and by fertilized mictic females (those producing resting eggs), $N_e(t)$ was calculated as:

$$4N_u(t-1)N_f(t)/(N_u(t-1) + N_f(t)) \tag{4}$$

(adapted from Hedrick 1984), where $N_u$ and $N_f$ are the abundance of unfertilized and fertilized mictic females, respectively. (Note that from the assumptions above, males fertilizing at $t$ were produced by unfertilized mictic females at $t-1$.) An appropriate population size per mating volume was used to transform population density into abundance. Mating volumes were obtained by calculating the volume from which mates are obtained by mictic females. At the end of the growing season, heterozygosity in the resting egg pool was estimated by averaging the heterozygosity of each resting egg.

## RESULTS

The simulated effects of a toxicant-caused reduction in population growth rate ($r$) during a growing season can be substantial (Fig. 3). In the model, $r$ values ranged from $-2.73$ to $2.80$ through different phases of population growth. Mictic reproduction

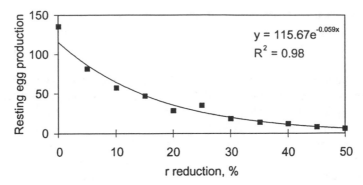

**FIG. 3.** Effects of reductions in population growth rate ($r$) on resting egg production in one growing season. Resting eggs are the number produced per liter; percent $r$ reduction is relative to a population lacking toxicant stress.

occurred on an average of 26 days per year in the simulated dynamics. Because the relationship between resting egg production and reduction in $r$ is a negative exponential ($y=115.7e^{-0.059x}$, $R^2=0.98$), small reductions in $r$ can lead to large reductions in resting egg production. For example, if toxicity reduces $r$ by 10%, resting egg production is reduced by about 50%. A 50% reduction in $r$ reduced resting egg production by about an order of magnitude. Toxicant-caused reductions in $r$ are the basis for the standard reproductive toxicity test with rotifers (Snell and Moffat 1992; Janssen et al. 1994), and such reductions in $r$ are typically one of the clearest responses of rotifer populations to toxicant stress.

The rate at which mictic females are fertilized has considerable effect on resting egg production (Fig. 4). Fertilization rate is a sigmoid function of population size,

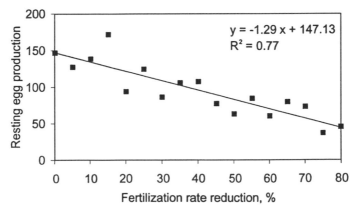

**FIG. 4.** Effects of reductions in fertilization rate on resting egg production in one growing season. Resting eggs are the number produced per liter; percent reduction in fertilization rate is relative to a population lacking toxicant stress.

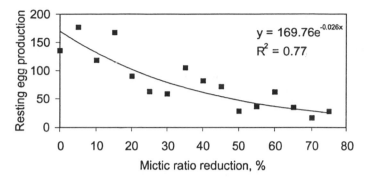

**FIG. 5.** Effects of reductions in mictic ratio on resting egg production in one growing season. Resting eggs are the number produced per liter; percent reduction in mictic ratio is relative to a population lacking toxicant stress.

mictic ratio, male density, and encounter rate. In the model, average fertilization rate was 13% and reductions up to 80% were explored. The effect of reduced fertilization on resting egg production was linear ($y = -1.29x + 147$, $R^2 = 0.77$). The scatter about the line is the result of the stochastic component of population growth built into the model. Initially resting egg production was about 150 per liter. A reduction in fertilization rate of about 65% cut resting egg production in half.

A potentially important effect of toxicants is on the proportion of daughters produced by a female that are mictic. This mictic ratio ranges from 8–69% in rotifer populations (Table 1) but is more typically about 25%, the ratio used in the model for conditions lacking toxicant stress. As the mictic ratio was reduced, resting egg production decreased exponentially ($y = 169.8e^{-0.026x}$, $R^2 = 0.77$) (Fig. 5). As an example, a reduction in mictic ratio of about 30% reduced resting egg production by about 50%. A 70% reduction in mictic ratio reduced resting egg production by nearly an order of magnitude. It is clear that if toxicity causes fewer sexual females to be produced due to a reduced mictic ratio, many fewer resting eggs will be produced by that population. In contrast to the large effects of mictic ratio, increases in mictic threshold had little effect on resting egg production (Fig. 6). Even five-fold mictic threshold increases from 5 to 25 females per liter only slightly decreased resting egg production.

Another long-term impact of toxicants on sexual reproduction in rotifers is loss of heterozygosity. The rate of heterozygosity loss is directly related to effective population size. Because toxicity reduces population growth rate ($r$) and therefore the size of a population during sexual reproduction, we investigated how reductions in $r$ increase the loss of heterozygosity (Fig. 7). In rotifers, effective population size ($N_e$) is determined by volume from which a mictic female obtains her mate multiplied by mictic female and male density. This volume depends on net dispersal distances of males and females (see Discussion). We explored several mating volumes for their effect on effective population size. For a volume of 100 liters, mean $N_e$ was 164. At this $N_e$, when $r$ is reduced from 0–50%, there was no effect on heterozygosity loss

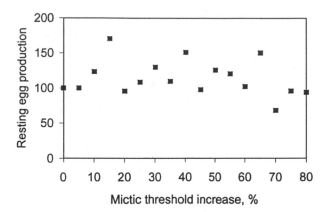

**FIG. 6.** Effects of reductions in mictic threshold on resting egg production in one growing season. Resting eggs are the number produced per liter; percent reduction in mictic threshold is relative to a population lacking toxicant stress.

over one growing season. Even at volumes of 10 liters (mean $N_e=18$), reductions in $r$ had little effect. However, at a volume of one liter (mean $N_e=2$), a 50% reduction in $r$ caused about a 25% reduction in heterozygosity.

Similar results were found for the effect of mictic ratio reductions on heterozygosity loss (Fig. 8). Reductions in mictic ratios of up to 50% had no effect on heterozygosity at volumes of 100 liters (mean $N_e=180$). Small effects were observed at volumes of 10 liters (mean $N_e=18$) and one liter (mean $N_e=2$). Even at one liter, a 50% reduction in mictic ratio reduced heterozygosity by only about 4%. As with resting egg

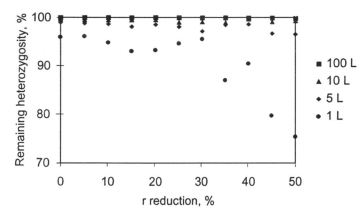

**FIG. 7.** Effects of mating volume and reductions in population growth rate ($r$) on heterozygosity. Volumes refer to the volume around the birth place of a sexual female from which her mate is obtained. Remaining heterozygosity refers to the average heterozygosity remaining in the resting egg pool at the end of one growing season.

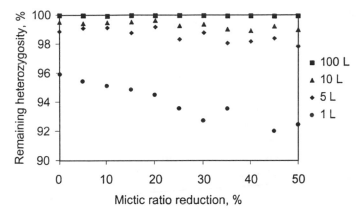

**FIG. 8.** Effects of mating volume and reductions in mictic ratio on heterozygosity. Volumes refer to the volume around the birth place of a sexual female from which her mate is obtained. Remaining heterozygosity refers to the average heterozygosity remaining in the resting egg pool at the end of one growing season.

production, mictic threshold increases had little effect on heterozygosity loss (data not shown). Even at volumes of one per liter, mictic threshold increases of 50% from an initial value of five mictic females per liter had no effect.

The effect of reductions in fertilization rate on heterozygosity loss were examined for a number of effective population sizes (Fig. 9). Reductions in fertilization rate of up to 50% had no effect on heterozygosity loss for volumes as small as five liters ($N_e=12$). Even when the mates of mictic females were derived from a volume of

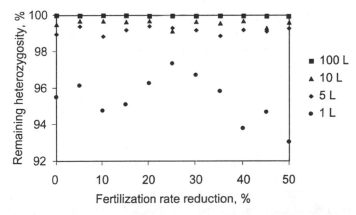

**FIG. 9.** Effects of mating volume and reductions in fertilization rate on heterozygosity. Volumes refer to the volume around the birth place of a sexual female from which her mate is obtained. Remaining heterozygosity refers to the average heterozygosity remaining in the resting egg pool at the end of one growing season.

only one liter around their point of birth ($N_e=2$), heterozygosity was not markedly reduced, fluctuating between a range of 94–97% of the initial heterozygosity.

## DISCUSSION

Work on rotifers has demonstrated that knowledge of toxicant effects on asexual reproduction is not necessarily predictive of effects on sexual reproduction (Snell and Carmona 1995). Differential effects of toxicants on sexual and asexual reproduction have been documented in other aquatic invertebrates. Intrapopulation variability was found in sublethal responses to cadmium stress in sexual and asexual gastropod populations of the family Hydrobiidae (Forbes et al. 1995). Cadmium reduced mean body growth rate in all the populations, but also increased variance in growth rate and altered the ranking of both means and variances among populations. The average intraclonal variability in growth in the parthenogenetic species was comparable with the intrapopulational variability found in the three closely related sexual species. Møller et al. (1996) observed lethal and sublethal responses of sexual and asexual gastropod species to cadmium. There were no differences in acute responses ($LC_{50}$s) between sexual and asexual species, but the asexual species had steeper concentration-response slopes (i.e., a more uniform response). In sublethal exposures, the asexual species was more sensitive to cadmium at low concentrations, but responses were similar at high cadmium concentrations. Although differences in sublethal responses to cadmium stress were demonstrated in closely related gastropod species, it is not known whether these differences were due to reproductive or species differences. This illustrates the advantage of using cyclical parthenogens like rotifers or cladocerans for investigating differential toxicant effects on sexual and asexual reproduction.

Results from the simulation suggest that toxicants can reduce resting egg production through effects on a variety of parameters, not all of which have equal impacts. Toxicants that reduce $r$ and mictic ratio tend to have the largest effects because these parameters are related to resting egg production as exponential functions. Toxicant-caused reductions in fertilization rate also reduced resting egg production, but this relationship is linear. Toxicants acting on the mictic threshold will not likely have an important impact on resting egg production. This is because most resting eggs are produced at a much higher density than the threshold population density, and both mictic female production and fertilization rate increase with population density. Consequently, even a five-fold increase in mictic threshold had only a marginal effect on resting egg production. Although we have been able to rank the relative impact of toxicants on some parameters, many other parameters remain unexplored (Fig. 2). The impact of toxicants on male rotifers, for example, has not been characterized. Little is known about the relative impact of toxicants on male survival, swimming behavior, mate recognition ability, and fertility. Experimental data on these parameters is currently inadequate for them to be incorporated into the model.

Mating patterns and resting egg production have very important influences on the genetic structure of zooplankton populations and their ability to adapt to environmental change (DeMeester 1996). The size and viability of the resting egg pool is critical to the long-term survival of most monogonont rotifer populations. Resting eggs can remain viable in aquatic sediments for decades (Marcus et al. 1994; Hairston et al. 1995), allowing zooplankton populations to survive hostile environments when reproduction may be impossible (Hairston 1996; Marcus 1996). This allows populations to endure several years of poor recruitment and to re-establish robust populations when environmental conditions become favorable. The size of the resting egg pool determines how fast rotifers can recolonize an environment and how long they can wait for favorable conditions to return. Furthermore, because resting eggs are produced sexually, hatching can release considerable genetic variation (Lynch and Gabriel 1983). What is not known is how much reduction in resting egg pool size is tolerable without jeopardizing the long-term survival of rotifer populations. The dynamics of resting eggs in natural sediments are virtually unknown. Examples of unexplored questions include: what percent of resting eggs remain in the top sediments and are available for hatching each year, what proportion usually hatch, what is typically the size of the founding population, and how does toxicant exposure affect hatching? Methods for measuring these effects are lacking even though human activities are increasingly impacting the distribution and abundance of natural zooplankton populations. Because current rotifer and cladoceran toxicity tests are based on asexual reproduction (e.g., Snell and Moffat 1992), they measure parameters that may have little impact on long-term population viability.

Heterozygosity is an important feature of populations because additive genetic variance determines the rate at which populations adapt to environmental stress (Hoffmann and Parsons 1991). Furthermore, losses of heterozygosity are likely to have a more permanent effect on a population than reductions in population abundance. Once lost, alleles can only be replaced by mutation or gene flow from another population. Heterozygosity is especially relevant in ecotoxicology because adaptation is a well-documented response to toxicant stress (e.g., Roush and McKenzie 1987) and is probably a key mechanism enabling zooplankton to tolerate toxicant exposure. Toxicant-caused directional selection and reductions in effective population size can have marked effects on heterozygosity. However, the relationship between toxicant exposure and heterozygosity is still poorly understood (Forbes 1996). Despite their theoretical importance, measures of genetic variability are not currently integrated into assessments of toxicity, nor is it clear how they might be.

The best estimates of heterozygosity in natural rotifer populations come from Gomez et al. (1995). Using protein electrophoresis, they estimated heterozygosity at four loci for four populations in Torreblanca Marsh on the Mediterranean coast of Spain. There are too few loci and populations to draw firm conclusions, but the average heterozygosity of 9% is a little lower than the median for invertebrates (Ward et al. 1992). This level of heterozygosity suggests that substantial genetic variability is present in these populations and that effective population size is not so small that most variation is eliminated by drift. Under the equilibrium heterozygosity model for neutral infinite alleles and mutation rates of $10^{-5}$ to $10^{-7}$ ($H_{eq}=4N_e\mu/(4N_e\mu+1)$), effective population size is 250–25,000.

In this paper, our approach was to examine how heterozygosity is lost as a result of reductions in effective population size. Simulating toxicant-caused reductions in parameters like $r$, mictic ratio, mictic threshold, and fertilization rate, we found that the loss of heterozygosity was minimal at effective population sizes above about 10. Heterozygosity loss was substantial at effective population sizes of about two when $r$ was reduced by 35–50%. Toxicant-simulated reductions in mictic ratio, mictic threshold, and fertilization rate, even at effective population sizes of about two, caused little loss of heterozygosity. Average effective population size was computed for all periods of sexual reproduction. However, most of the resting eggs were produced at quite large population sizes, as resting egg production is positively density-dependent. This density-dependence restricts sexual reproduction to large population sizes, preventing heterozygosity loss due to sampling errors. The existence of a density threshold for sexual reproduction also helps to avoid sex at small population sizes when the effects of drift are strongest.

The analysis of heterozygosity loss raises the question of what are reasonable effective population sizes for natural rotifer populations? This question is more complicated than it may seem because of the biological features of sexual reproduction in rotifers. When a sexual female *Asplanchna* is born, she has about four hours to be fertilized (Buchner et al. 1967). After that, even if impregnated, females will produce haploid males. Consequently, a sexual female's mate will come from a relatively small volume of water around her as compared to the entire volume of the lake. This isolation by distance has been described by King and Murtaugh (1997) in a neighborhood model modified from Wright (1969). They calculated that a sexual female's mate comes from a sphere of about one meter diameter, which is determined by female and male swimming speeds and net dispersal distances, assuming movement in a straight line. Such a sphere has a volume greater than 1000 liters and is even larger if there are intervening parthenogenetic generations. Our examination of this question suggests that this estimate is probably about two orders of magnitude too high. This results from their assumption that females swim in the same direction for their entire four-hour fertile period. We have modeled net dispersal distance in *Asplanchna girodi* as a diffusion process and found that mates could come from a volume as small as 1–10 liters around the point of birth of a sexual female. Effective population sizes of natural *Asplanchna* populations, therefore, may be much smaller than previously believed. This also illustrates the principal contribution of this work—identifying variables like effective population size that have not been considered in ecotoxicological studies. Subtle effects of toxicants on parameters like net dispersal distance could have large effects on effective population size and, therefore, the rate of heterozygosity loss.

There are several implications of our results for ecotoxicology. First, the asexual reproduction currently being measured by standard toxicity tests may not be very relevant for the long-term survival of cladoceran and rotifer populations. The relationship between resting egg pool size, genetic variability, and the evolutionary fate of zooplankton populations needs further elaboration. It would be useful to develop methods incorporating measures of resting egg pool size and viability into toxicity assessments. Likewise, it would be advantageous to know the relative impact of a variety of parameters on effective population size. Better understanding is needed of how toxicants affect parameters determining effective population size like male and female swimming speed, net dispersal distance, and mating probability. By emphasizing the development of convenient, rapid toxicity tests, ecotoxicologists may have overlooked variables that have the greatest impact on the evolutionary fate of zooplankton populations. Improved knowledge of these variables will permit the incorporation of measurements into toxicity tests that more accurately assess the long-term impact of toxicants without necessarily increasing effort.

## ACKNOWLEDGMENTS

This work was supported in part by grant INT-9424356 from the National Science Foundation and by grant SAB95-0219 from the Spanish Ministerio de Educacion y Ciencia.

## REFERENCES

Bell, G. 1982. *The masterpiece of nature: The evolution and genetics of sexuality.* Berkeley: University of California Press.
Birky, Jr., C. W. and J. J. Gilbert. 1971. Parthenogenesis in rotifers: The control of sexual and asexual reproduction. *American Zoologist* 11:245–266.
Buchner, H., C. Mutschler, and H. Kiechle. 1967. Die determination der manchen- und dauereiproducktion bie *Asplanchna sieboldi*. *Biologisches Zentralblatt* 86:599–621.
Carlin, B. 1943. Die planktonrotatorien des motalastrom. *Meddelanden frun Lunds Universitet Limnologiska Institution* 17:1–163.
Carmona, M. J., A. Gomez, and M. Serra. 1995. Mictic patterns of the rotifer *Brachionus plicatilis* in small ponds. *Hydrobiologia* 313/314:365–371.
Carmona, M. J., M. Serra, and M. R. Miracle. 1994. Effect of population density and genotype on life-history traits in the rotifer *Brachionus plicatilis* O F Muller. *Journal of Experimental Marine Biology and Ecology* 182:223–235.
Crow., J. F., and M. Kimura. 1970. *An introduction to population genetics theory.* New York: Harper and Row.
DeMeester, L. 1993. Inbreeding and outbreeding depression in *Daphnia*. *Oecologia* 96:80–84.
DeMeester, L. 1996. Local genetic differentiation and adaptation in freshwater zooplankton populations: Patterns and processes. *Ecoscience* 4:385–399.
Depledge, M. H. 1994. Genotypic toxicity: Implications for individuals and populations. *Environmental Health Perspectives* 102(suppl 12):101–104.
Depledge, M. H. 1996. Genetic ecotoxicology: An overview. *Journal of Experimental Marine Biology and Ecology* 200:57–66.
Fisher, R. A. 1930. *The genetical theory of natural selection.* Oxford: Clarendon.
Forbes, V. E. 1996. Chemical stress and genetic variability in invertebrate populations. *Toxicology and Ecotoxicology News* 3:136–141.
Forbes, V. E., V. Møller, and M. H. Depledge. 1995. Interindividual variability in sublethal response to heavy metal stress in sexual and asexual gastropod populations. *Functional Ecology* 9:477–84.
Gerritsen, J. 1980. Sex and parthenogenesis in sparse populations. *American Naturalist.* 115:718–742.

Gilbert, J. J., and J. R. Litton. 1978. Sexual reproduction in the rotifer *Asplanchna girodi*: Effects of tocopherol and population density. *Journal of Experimental Zoology* 204:113–122.

Gilbert, J. J. 1975. Polymorphism and sexuality in the rotifer *Asplanchna* with special reference to the effects of prey-type and clonal variation. *Archives für Hydrobiologia* 75:442–483.

Gomez, A., M. Temprano, and M. Serra. 1995. Ecological genetics of a cyclical parthenogen in temporary habitats. *Journal of Evolutionary Biology* 8:601–602.

Gomez, A. 1996. Ecologia genetica y sistemas de reconocimiento de pareja en poblaciones simpatricas de rotiferos. Ph.D. Thesis, Valencia, Spain: University of Valencia.

Gomez, A., M. J. Carmona, and M. Serra. 1997. Ecological factors affecting gene flow in the *Brachionus plicatilis* complex. *Oecologia* 111:350–356.

Grant, A., J. G. Hateley, N. V. Jones. 1989. Mapping the ecological impact of heavy metals on the estuarine polychaete *Nereis diversicolor* using inherited metal tolerance. *Marine Pollution Bulletin* 20:235–238.

Guttman, S. I. 1994. Population genetic structure and ecotoxicology. *Environmental Health Perspectives* 102(suppl 12):97–100.

Hagiwara, A., and A. Hino. 1989. Effect of incubation and preservation on resting egg hatching and mixis in the derived clones of the rotifer *Brachionus plicatilis*. *Hydrobiologia* 186/187:415–421.

Hairston, Jr. N. G., R. A. van Brunt, and C. M. Kearns. 1995. Age and survivorship of diapausing eggs in a sediment egg bank. *Ecology* 76(6):1706–1711.

Hairston, Jr. N. G. 1996. Zooplankton egg banks as biotic reservoirs in changing environments. *Limnology and Oceanography* 41(5):1087–1092.

Hedrick, P. 1984. *Population biology*. Boston, MA:Jones and Bartlett Publishers.

Hoffmann, A. A., and P. A. Parsons. 1991. *Evolutionary genetics and environmental stress*. New York: Oxford University Press.

Innes, D. J. 1989. Genetics of *Daphnia obtusa*: Genetic load and linkage analysis in a cyclical parthenogen. *Journal of Heredity* 80:6–10.

Janssen, C. R., G. Persoone, and T. W. Snell. 1994. Cyst-based toxicity tests. VIII Short-chronic toxicity tests with the freshwater rotifer *Brachionus calyciflorus*. *Aquatic Toxicology* 28:243–258.

King, C. E. 1970. Comparative survivorship and fecundity of mictic and amictic female rotifers. *Physiological Zoology* 43:206–212.

King, C. E. 1993. Random genetic drift during cyclical ameiotic parthenogenesis. *Hydrobiologia* 255/256:205–212.

King, C. E., and P. Murtaugh. 1997. Effects of asexual reproduction on the neighborhood area of cyclical parthenogens. *Hydrobiologia* 358:55–62.

King, C. E., and T. W. Snell. 1980. Density dependent sexual reproduction in natural populations of the rotifer *Asplanchna girodi*. *Hydrobiologia* 73:149–152.

Kleiven, O. T., P. Larsson, and A. Hoback. 1992. Sexual reproduction in *Daphnia magna* requires three stimuli. *Oikos* 65:197–206.

Klerks, P. L., and J. S. Levinton. 1989. Rapid evolution of metal resistance in a benthic oligochaete inhabiting a metal-polluted site. *Biological Bulletin* 176:135–141.

Lynch, M., and W. Gabriel. 1983. Phenotypic evolution and parthenogenesis. *American Naturalist* 122:745–764.

Lynch, M., and H. W. Deng. 1994. Genetic slippage in response to sex. *American Naturalist* 144:242–261.

Marcus, N. H. 1996. Ecological and evolutionary significance of resting eggs in marine copepods: Past, present and future. *Hydrobiologia* 320:141–152.

Marcus, N. M., R. Lutz, W. Burnett, and P. Cable. 1994. Age, viability and vertical distribution of zooplankton resting eggs from an anoxic basin: Evidence of an egg bank. *Limnology and Oceanography* 39:154–158.

Maroni, G., J. Wise, J. E. Young, and E. Otto. 1987. Metallothionein gene duplications and metal tolerance in natural populations of *Drosophila melanogaster*. *Genetics* 117:739–744.

Mort, M. A. 1991. Bridging the gap between ecology and genetics. *Trends in Ecology and Evolution* 6:41–45.

Møller, V., V. E. Forbes, and M. H. Depledge. 1996. Population responses to acute and chronic cadmium exposure in sexual and asexual estuarine gastropods. *Ecotoxicology* 5:313–326.

Pourriot, R., and C. Rougier. 1977. Effects de la densite de population et groupement sur la reproduction de *Brachionus calyciflorus* (Pallas) (Rotifera). *Annales de Limnologie* 13:101–113.

Pourriot, R., and C. Rougier. 1979. Influences conjuguies du groupement et de la qualiti de la nourriture sur la reproduction de *Brachionus plicatilis* O F Muller (Rotifera). *Netherlands Journal of Zoology* 19:242–264.

Roush, R. T., and J. A. McKenzie. 1987. Ecological genetics of insecticide and acaricide resistance. *Annual Review of Entomology* 32:361–380.

Ruttner, F. 1930. Das plankton des lunzer untersees. *Internationale Review Gesamten Hydrobiologie* 23:1–287.

Ruttner-Kolisko, A. 1974. Plankton rotifers: Biology and taxonomy. *Binnengewasser* 26:1–146.

Ruttner-Kolisko, A. 1985. Results of individual cross-mating experiments in three distinct strains of *Brachionus plicatilis* (Rotatoria). *Verfahren Internationale Verein Limnologie* 22:2979–2981.

Serra, M. 1987. Variacion morfomitrica isoenzimatica y demografica en poblaciones de *Brachionus plicatilis*: diferenciacion genitica y plasticidad fenotipica. Ph.D. Thesis, Valencia, Spain: Valencia University.

Snell, T. W. 1980. Blue-green algae and selection in rotifer populations. *Oecologia* 46:343–346.

Snell, T. W., and B. L. Garman. 1986. Encounter probabilities between male and female rotifers. *Journal of Experimental Marine Biology and Ecology* 97:221–230.

Snell, T. W., and M. Childress. 1987. Aging and loss of fertility in male and female *Brachionus plicatilis* (Rotifera). *International Journal of Invertebrate Reproduction and Development* 12:103–110.

Snell, T. W., and E. M. Boyer. 1988. Thresholds for mictic-female production in the rotifer *Brachionus plicatilis* (Muller). *Journal of Experimental Marine Biology and Ecology* 124:73–85.

Snell, T. W., and B. D. Moffat. 1992. A two day life cycle test with the rotifer *Brachionus calyciflorus*. *Environmental Toxicology and Chemistry* 11:1249–1257.

Snell, T. W., and M. J. Carmona. 1995. Comparative sensitivity of sexual and asexual reproduction in the rotifer *Brachionus calyciflorus*. *Environmental Toxicology and Chemistry* 14:415–420.

Snell, T. W., and M. Serra. 1998. Dynamics of natural rotifer populations. *Hydrobiologia* (in press).

Wallace, R. L., and T. W. Snell. 1991. *Rotifera. Classification of North American freshwater invertebrates*, 187–248. New York: Academic Press.

Ward, R. D., D. O. F. Skibinski, and M. Woodwark. 1992. Protein heterozygosity, protein structure and taxonomic differentiation. *Evolutionary Biology* 26:73–159.

Wright, S. 1969. *Evolution and the genetics of populations. Vol. 2. The theory of gene frequencies.* Chicago: University of Chicago Press.

Wright, S. 1977. *Evolution and the genetics of populations. Vol. 3. Experimental results and evolutionary deductions.* Chicago: University of Chicago Press.

*Genetics and Ecotoxicology*
Edited by V. E. Forbes
Copyright © 1999 Taylor & Francis

# 10

# The Influence of Reproductive Mode and Its Genetic Consequences on the Responses of Populations to Toxicants: A Case Study

Valery E. Forbes, Vibeke Møller, Robert A. Browne,
and Michael H. Depledge

**Abstract.** Uniformity in response of test populations has long been an important consideration in the development of toxicity and ecotoxicity tests. We employed the brine shrimp *Artemia* as a model organism to examine the influence of reproductive mode and its genetic consequences on the response uniformity of test populations to toxicant exposure. We measured growth in four obligately parthenogenetic populations (*Artemia parthenogenetica*) and three sexual populations (two *A. franciscana* and one *A. urmiana*) at three copper concentrations. For parthenogenetic populations, offspring from several females were tested whereas full-sib offspring from single mating pairs were used for sexual populations. We found that *Artemia* derived from geographically distinct populations of the same species responded similarly to copper exposure. Although there were significant differences in the shape of the concentration-response curves among the three species tested, these differences were not related to reproductive mode. Interactions between clone and copper treatment were a significant source of phenotypic variability for the parthenogenetic populations and accounted for up to 20% of the total phenotypic variance. Residual variability among individuals (within-clones) accounted for the largest percent of the total phenotypic variability, ranging between 55 and 84%. Different clones within parthenogenetic populations differed as much from each other in their response to copper as did different families within the sexual populations. Furthermore, within-treatment random variation among individuals was generally not significantly lower within single clones than within families of genetically mixed offspring. The results of this case study cast doubt on the notion that the employment of genetically uniform populations offers a substantial advantage in terms of increasing the phenotypic homogeneity of a test population's response to toxicant exposure. The primary advantage of asexual populations is that they can facilitate the identification and quantification of important sources of variability in phenotypic responses to toxicants—an issue that remains in need of study.

**Keywords.** *Artemia*; copper; geographical variability; parthenogenesis; phenotypic variability.

## INTRODUCTION

Clonal organisms and inbred strains of sexual species are widely used in toxicological and ecotoxicological testing and risk assessment schemes for the purpose of minimizing variability in test results (Gaddum 1933; OECD 1997). Implicit in this approach is the assumption that genetic differences among individuals in a test population are an important source of within-species phenotypic variability in toxicant responses. This assumption is widely made despite the absence of data showing that genetic uniformity is indeed expressed as phenotypic uniformity in the responses of organisms to toxicants (Forbes and Depledge 1993, 1996; Forbes and Forbes 1994). In this context the performance of asexual, relative to sexual, populations and the variability in sensitivity to toxicant exposure within and among genotypes is a topic of great concern (Baird et al. 1991; Baird 1992, 1993; Forbes and Depledge 1993, 1996; Baird and Barata chapter 11).

Here we provide the results of a case study in which growth rates of sexual and asexual (parthenogenetic) populations of the brine shrimp *Artemia* were examined after being exposed to copper. We selected *Artemia* as a model organism because 1) the genus consists of several related species that reproduce either asexually (by parthenogenesis) or sexually but not both, 2) it has a geographic distribution that encompasses Europe, America, and Asia, and 3) its life-history characteristics (e.g., short lifespan, small body size, etc.) make it amenable for laboratory ecotoxicity tests.

We selected copper because, although it is an essential metal, it is highly toxic at the elevated concentrations that may arise from anthropogenic inputs (Sunda and Hanson 1987).

We measured growth rate because it is frequently used as a sublethal endpoint in ecotoxicological studies, and the energy available for growth (or for growth and reproduction once individuals become sexually mature) is assumed to be correlated with fitness (Koehn and Bayne 1989). In juveniles, the rate of growth influences the age and size at reproductive maturity, and these traits can have an important influence on population growth rate (Stearns 1992). Stress from toxicant exposure may cause a reduction in net energy balance and production (Koehn and Bayne 1989; Willows 1994) and/or cause available energy to be reallocated away from growth and toward metabolic expenditures involved in detoxification (Forbes and Calow 1996).

We examined growth in response to copper exposure in three *Artemia* species from a total of seven geographically distinct populations as follows: 1) we compared growth between sexual and parthenogenetic species; 2) we compared growth among geographically distinct populations within species; 3) we compared growth among families (among clones or among full-sib families) within populations; 4) we compared growth among individuals within a single clone or full-sib family. We used these comparisons to ask the following questions:

1. Are there substantial differences in the sensitivity of parthenogenetic versus sexual species to copper exposure? Large differences in sensitivity to toxicants as a

consequence of reproductive mode could lead to biased test results that are either over- or under-conservative (e.g., Snell and Carmona 1995; Snell et al. chapter 9).
2. Do populations of the same species collected from geographically distinct areas differ appreciably in their response to copper exposure? Although all of the species used here were maintained under identical laboratory conditions for several generations before copper exposure, differences in selection pressures in their source habitats could potentially influence their performance in ecotoxicity tests.
3. Are genotype × environment interactions a significant source of phenotypic variability? In other words, does the effect of copper on growth rate differ significantly among clones (or full-sib families) within a population? If all clones show similar concentration-response curves, then one clone may adequately represent the response of the species. However, if the form of the concentration-response curve differs substantially among clones, this would argue for using a suite of clones for testing purposes.
4. Is the average phenotypic variability within clones less than the average phenotypic variability within full-sib families? If it is, this suggests that genetic differences among siblings are of measurable importance in controlling phenotypic responses to toxicants. If the average variances for clones and siblings are similar, it suggests that the influences of microenvironmental heterogeneity and/or developmental noise (rather than genotype) are important in controlling phenotypic variability within a test population and that the use of clones is not likely to be an effective way to reduce variability in ecotoxicological test results. If the average variances for clones are greater than those for sexual siblings, it may indicate that clonal genotypes are less phenotypically stable than their sexual counterparts (i.e., are more sensitive to unmeasurable environmental heterogeneity and/or are developmentally unstable). If this were the case, the employment of clonal populations could lead to an *increase* in the variability of test results, and this would argue strongly against the use of clonal genotypes for standardized ecotoxicological testing protocols.

## CASE STUDY

### Study Species

*Artemia* is found in areas of solar salt production, at salinities above 70‰, where high salt levels inhibit predators. Most species primarily reproduce oviviparously (nauplii), but some may also produce encapsulated embryos (cysts) that undergo diapause and that are highly resistant to environmental stress. *Artemia* is thus restricted in its distribution to a limited range of ecological conditions, but within its range it is extremely successful. Some species reproduce sexually whereas others are obligate parthenogens. Only sexually reproducing species are found in the New World, whereas both sexual and parthenogenetic species are found in the Old World. The genus *Artemia* is therefore split into three major groups including one or more species:

**TABLE 1.** *Reproductive mode, population code, geographic origin, species designation and ploidy of the different* Artemia *populations*

| Code | Geographic origin | Species name | Ploidy |
|---|---|---|---|
| Parthenogenetic populations (PAR) | | | |
| KU | Kutch, India | *A. parthenogenetica* | triploid |
| TS | Tientsin, China | *A. parthenogenetica* | diploid |
| MS | Margherita di Savoia, Spain | *A. parthenogenetica* | diploid |
| SG | Salin de Giraud, France | *A. parthenogenetica* | diploid |
| New world sexual populations (NWS) | | | |
| SF | San Francisco Bay, CA, USA | *A. franciscana* | diploid |
| GS | Great Salt Lake, UT, USA | *A. franciscana* | diploid |
| Old world sexual populations (OWS) | | | |
| YC | Yuncheng, Shanxi Province, China | *A. urmiana* | diploid |

New World sexual (NWS) populations (*Artemia persimillis, Artemia monica,* and *Artemia franciscana*), Old World sexual (OWS) populations (*Artemia tunisiana* and *Artemia urmiana*) and parthenogenetic (PAR) populations (*Artemia parthenogenetica*) (Browne and Bowen 1991). Although species can be distinguished morphologically, the anatomical variability displayed by *Artemia* is quite small, and the species have been considered to be sibling species (Abreu-Grobois and Beardmore 1982).

The geographic origin, species designation, and ploidy of the *Artemia* populations used in this study are listed in Table 1.

All sexual populations are diploid, whereas some parthenogenetic populations are diploid and others are polyploid. Heterozygosity has been shown to increase with increasing ploidy in the parthenogenetic populations and is usually high in sexual populations, especially in the OWS population YC. Polyploid parthenogenetic populations are essentially monoclonal whereas diploid populations may consist of a mixture of clones (Abreu-Grobois and Beardmore 1982; Abreu-Grobois 1987). Comparisons of genetic distances (Nei's D) show extensive genetic divergence between Old World and New World species. The closest relative to the parthenogenetic species is the OWS species, *Artemia urmiana* (Abreu-Grobois and Beardmore 1982; Abreu-Grobois 1987; Pilla and Beardmore 1994).

## Experimental Design

For the sexual populations, test pairs were obtained by mass hatching of cysts (originating from natural populations) in 35‰ artificial seawater (35 g/l Instant Ocean in distilled water). Upon hatching nauplii were transferred to 2l jars containing 90‰ artificial seawater (35 g/l Instant Ocean and 55 g/l salt, NaCl, added to distilled water) and reared until sexual maturity. Every fourth day the contents of a jar were split into two jars and fresh brine was added to avoid density limitation of growth. Due to low encystment rates, nauplii from the parthenogenetic populations were obtained from reproducing females that were kept individually. These nauplii were raised at the same densities (adjusting jar size and water volume accordingly) as the sexual nauplii.

Throughout the experiment, individuals were fed a diet of mixed algae and dried yeast supplied at the maximum rate that could be cleared each day. Food supply was therefore presumed not to be a limiting factor. All cultures were maintained under 24 hour fluorescent light at room temperature, 20–25°C.

When sexually mature, individual females (parthenogenetic) and male-female pairs (sexual) were separated and kept in 300 ml jars. Nauplii recovered from individual parthenogenetic females are thus genetically identical (single clones). Nauplii from pairs of sexual individuals are defined as a full-sib family. The sexual populations are represented by a random sample of families (four to eight families per population). Parthenogenetic populations had a lower reproductive output (with fewer females breeding), so that only two to five clones per population were available. For MS and KU, the clones were independently derived whereas two of the clones from both TS and SG had the same grandmother. Allozyme analyses of surviving mothers or siblings were performed (ADH, G-6PDH, $\alpha$-GPDH, PEP, SDH, PGM, EST (3 loci), SOD, and LDH), and these showed substantial differences among the parthenogenetic populations, but only one population (SG) exhibited allozyme differences among clones (i.e., one of the clones differed from the other four).

Sibling nauplii were kept in mass cultures (approximately 10 nauplii per 100 ml brine) in control conditions for two days before starting the growth experiment. If a minimum of 24 nauplii (from a clone/family) were available after two days they were randomly assigned to one of three treatments: control, 50 $\mu$g/l Cu, and 100 $\mu$g/l Cu. Eight nauplii per replicate were placed in a jar containing 200 ml brine (90‰) of the appropriate copper concentration (Table 2). After six days survival was checked, and if mortality had occurred in one treatment group the number of nauplii in the other treatment groups was adjusted to ensure equal densities. At the same time excess algae, yeast, and accumulated feces were removed by pipette, and 100 ml fresh brine was added. After 12 days the surviving nauplii were frozen (−80°C) following narcotization with carbonated water. Length was measured using an image analysis system, and length at 14 days after birth was used as a measure of growth. The initial size of nauplii could not be measured with this system, however differences in starting size among populations were minor relative to the large differences in final size (Møller, personal observation). Because we are primarily concerned with comparing relative contributions of different sources of variability to growth and not absolute growth rates among species, small differences in starting size among species or populations were of minor concern (cf. Baird and Barata, chapter 11).

## Statistical Analysis

We analyzed growth using a mixed model ANOVA design and partitioned the total variance in growth to variance 1) among species, 2) among populations, within species, 3) among clones (for PAR) or full-sib families (NWS, OWS) within populations, and 4) within clones or full-sib families. Main factors in the ANOVA were population (considered random) and copper concentration (considered fixed). The

**TABLE 2.** *Number of clones/families from each population used in the growth experiment, numbers assigned to each treatment**

| Code | Clones/ families | Treatment Control | 50 ppb | 100 ppb | N |
|------|------------------|---------|--------|---------|---|
| Parthenogenetic populations (PAR) | | | | | |
| KU | 5 | 3 (14) | 2 (7) | 1 (3) | 24 |
| | | 3 (9) | 5 (25) | 5 (22) | 56 |
| | | 4 (27) | 5 (32) | 3 (21) | 80 |
| | | 4 (27) | 5 (27) | 5 (29) | 83 |
| | | 8 (37) | 6 (27) | 5 (20) | 84 |
| TS | 3 | 1 (6) | 1 (5) | 1 (4) | 15 |
| | | 1 (8) | 1 (8) | 1 (3) | 19 |
| | | 2 (8) | 2 (7) | 2 (11) | 26 |
| MS | 2 | 1 (8) | 1 (8) | 1 (3) | 19 |
| | | 6 (28) | 5 (29) | 5 (28) | 85 |
| SG | 4 | 3 (19) | 3 (16) | 2 (11) | 46 |
| | | 5 (35) | 4 (25) | 6 (23) | 93 |
| | | 6 (36) | 6 (40) | 5 (27) | 103 |
| | | 5 (34) | 6 (45) | 5 (39) | 118 |
| New world sexual populations (NWS) | | | | | |
| SF | 5 | 1 (3) | 1 (4) | 1 (6) | 13 |
| | | 2 (6) | 1 (2) | 2 (6) | 14 |
| | | 1 (5) | 1 (6) | 1 (4) | 15 |
| | | 4 (16) | 5 (20) | 5 (16) | 52 |
| | | 3 (22) | 4 (27) | 4 (27) | 63 |
| GS | 8 | 1 (7) | 1 (7) | 1 (5) | 19 |
| | | 1 (6) | 1 (7) | 1 (7) | 20 |
| | | 1 (7) | 2 (14) | 1 (6) | 27 |
| | | 2 (11) | 1 (6) | 2 (11) | 28 |
| | | 2 (15) | 2 (14) | 1 (6) | 35 |
| | | 2 (15) | 2 (16) | 1 (4) | 35 |
| | | 3 (19) | 2 (6) | 3 (15) | 40 |
| | | 6 (43) | 6 (41) | 5 (29) | 113 |
| Old world sexual populations (OWS) | | | | | |
| YC | 4 | 3 (16) | 3 (18) | 3 (19) | 53 |
| | | 3 (17) | 3 (19) | 4 (28) | 64 |
| | | 4 (24) | 6 (29) | 5 (30) | 83 |
| | | 8 (39) | 6 (34) | 5 (32) | 105 |

*First number: number of replicate jars, second number: the final number of nauplii, and N: total number of nauplii from each clone/family.

model also included interaction between main factors as well as an underlying nested design in which populations were nested within species and clones/families were nested within populations. Mixed model F-tests were constructed using expected mean squares according to Scheffés model (Fry 1992) which can be used for un-balanced designs. In mixed models the F-test for a fixed factor is only approximate (Winer 1971). Significant main factors indicate differences in marginal means (Fry 1992) (e.g., a significant clone/family term would indicate within-population genetic differences in average growth over all copper treatments, whereas a significant clone/family × concentration interaction would indicate differences in sensitivity of response to copper among clones/full-sib families). Nauplii densities (for day 1–6 and 6–12) were initially tested as covariates but were not significant and were subsequently omitted from further analysis. Significant factors were compared using Bonferroni's adjusted pairwise comparisons of means compensating for the number of nonindependent tests performed (Kirby 1993) to test for differences among species, populations, and clones/full-sib families. In addition, specific questions about treatment effects were tested using linear contrasts or response surfaces (Kirby 1993). Separate mixed-model ANOVAs were performed for each population to examine within-population patterns of genetic variation for sensitivity.

The underlying assumptions of ANOVA (normality and homogeneity of group variances) were checked by visual inspection of residual plots. Small groups tended to have less variation, but in general the homogeneity of group variances assumption was met, and the data were reasonably normal. ANOVA is robust to small deviations from assumptions and is the only analysis that allows testing of nested designs and derivation of variance components. Furthermore, the ANOVA results were compared to weighted means analysis and nonparametric ANOVAs to confirm the results.

Two methods were used to partition variance components:

### Variance Partitioning From ANOVA

Interpretations of F-tests from ANOVAs are very dependent on the degrees of freedom (df). For instance, the large number of error df in the models tested in this study makes nearly all interactions significant. The total phenotypic variance can be partitioned into different components by analyzing a series of separate mixed models (by population) and solving a system of simultaneous equations prescribed by the calculated mean squares and their expectations. Specifically,

$$s_P^2 = s_G^2 + s_E^2 + s_{G \times E}^2 + s_R^2$$

where $s_P^2$, the total phenotypic variation, is split into genetic variation $s_G^2$ (among populations or clones/full-sib families), environmental variance $s_E^2$ (differences due to Cu-treatments) an interaction term $s_{G \times E}^2$ (genotype by Cu-treatment) and residual $s_R^2$ (differences among individuals of a single population or clone/full-sib family in one Cu-treatment). For the clonal replicates, the residual variation includes effects caused by developmental noise, measurement error and microenvironmental effects. For the sexual populations, the full-sibling are not genetically identical, and therefore some of the genetic variance among siblings is included in the residual term. This may underestimate the genetic variance ($s_G^2$) as well as possibly the genotype by treatment variance ($s_{G \times E}^2$). Nevertheless, comparing these variance terms between single clones and genetically variable siblings can provide insight into the importance of genetic versus random factors as sources of phenotypic variability (see question 4, Introduction).

Occasionally the partitioning of variance components results in negative estimates of some of the terms (Winer 1971). Because the cases for which we obtained negative values always accounted for less than 3% of the total variation, they were treated as zero, and subsequently the other variance components were recalculated. The partitioning was performed following procedures given in Milliken and Johnson (1992). It was not possible to calculate standard errors for variance components as these are unknown for complicated unbalanced designs (Sokal and Rohlf 1996).

## Coefficient of Variation

Differences in variation among species and treatments can be compared using the coefficient of variation CV, which is a measure of variation often employed when the mean and standard deviation are positively correlated (which is the case for many biological variables) (Sokal and Rohlf 1996). Our replication within clones/full-sib families allowed for calculations of within-treatment coefficients of variation for each treatment and clone/full-sib family cell ($=CV_{Within}$). These were corrected for different sample sizes following Sokal and Rohlf (1996). We then performed a combined factorial-hierarchical ANOVA to test differences in CVs in the same way as we tested differences among means. In addition, CVs across copper treatments ($CV_{Across}$) for each of the clones/full-sib families were calculated. The averages for all clones/full-sib families within each population were compared using Kruskal-Wallis nonparametric ANOVA, as these CVs were not normally or homogeneously distributed. These provided a measure of the average sensitivity of the seven populations to copper exposure.

The residual variance component of the variance partitioning approach ($s_R^2$) is comparable to the within-treatment coefficient of variation ($CV_{Within}$) for single clones or full-sib families. The across-treatment coefficient of variation ($CV_{Across}$) is often used as a measure of sensitivity, i.e., phenotypic plasticity (Bierzychudek 1989; Weider 1993). It is comparable to the environmental variance ($s_E^2$) combined with the genotype $\times$ environment interaction ($s_{G \times E}^2$) variance component (Dutilleul and Potvin 1995). In studies of selection on phenotypic plasticity some authors use this combined measure as the plasticity (e.g., Scheiner and Goodnight 1984; Yampolski and Scheiner 1994) whereas others take a more conservative approach and use only the interaction term (e.g., Via 1984; Weis and Gorman 1990). We have reported the terms separately, but also discuss the combined values.

The relationship between average growth and the sensitivity of growth rate to copper was investigated by a correlation analysis of population mean growth (averaged across copper treatments) with the combined sensitivity term derived from the variance partitioning ($s_{G \times E}^2 + s_E^2$) (Michaels and Bazzaz 1989). A positive correlation between average growth and sensitivity indicates that populations with the highest growth rates are the most sensitive (show the greatest reductions in growth rate) upon exposure to copper, whereas the slowest growing clones are the least sensitive to copper exposure.

Statistical significance is defined as $p \leq 0.05$; $p$-values $0.05 < p < 0.1$ are considered marginally significant. Statistical analyses were performed with Systat, version 5.04 (Wilkinson et al. 1992).

## RESULTS

### Average Growth

#### *Overall Differences*

The overall analysis of variance (Table 3) showed significant negative effects of copper treatment on growth, significant differences among the seven populations, but only marginally significant differences among populations within each of the three species. The same pattern was shown for interaction effects. That is, there was a significant overall interaction between treatment and population (among all seven populations), but within each species there was no interaction between treatment and population. Growth rates of *Artemia* that were derived from geographically distinct populations of the same species responded similarly to copper exposure, whereas there were significant differences in the shape of the concentration-response curves among the three species. In addition, there were significant differences among clones/full-sib families within populations and significant interactions between concentration and clones/full-sib families.

The remaining results are described sequentially beginning at the highest level of analysis.

#### *Differences Among Species*

Figure 1 summarizes the results of growth versus copper treatment at the species level. Even though there was no significant difference among species (main effect), there was a significant interaction between species and concentration. The most obvious

**TABLE 3.** *Overall analysis of variance*[†]

| Source | df | MS | M# | F-ratio | F |
|---|---|---|---|---|---|
| Treatment | 2 | 18.86 | M1 | M1/M6 | 4.75** |
| Population (model) | 6 | 36.96 | M2 | M2/M6 | 9.31** |
| Species | 2 | 44.54 | M3 | M3/M4 | 1.59 |
| Population w. species | 4 | 27.93 | M4 | M4/M5 | 2.33* |
| Clone/family w. population | 24 | 11.99 | M5 | M5/M10 | 15.57** |
| Treatment × Pop. (model) | 12 | 3.97 | M6 | M6/M10 | 5.16** |
| Treatment × Species | 4 | 8.13 | M7 | M7/M8 | 4.31** |
| Tmt. × Pop. w. species | 8 | 1.89 | M8 | M8/M9 | 0.70 |
| Tmt. × Clone/family w. pop. | 48 | 2.69 | M9 | M9/M10 | 3.49** |
| Error | 1537 | 0.77 | M10 | | |

*$p < 0.1$.

**$p < 0.05$.

[†]Treatment (Tmt.), population (Pop.), and their interactions constitute the overall model. These are further divided into species, populations nested within (w.) species, and clones/families nested within populations, for main effects as well as for interactions.

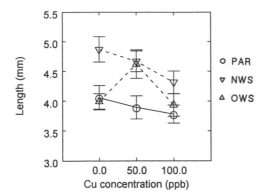

**FIG. 1.** Mean length of the different species as a function of copper concentration. Error bars are standard errors.

difference was the increased length at the intermediate concentration in YC, the OWS population. The NWS species decreased in length with increasing Cu concentration, whereas the parthenogenetic species was not significantly affected by Cu exposure.

### Differences Among Populations

Figure 2 shows the mean length of the different populations plotted against copper concentration. Differences among populations within species were only marginally significant. Group comparisons using linear contrasts showed that the two PAR populations, MS and SG, were significantly different, as were the two NWS populations, GS and SF. For the parthenogenetic populations there was neither an effect of treatment nor an interaction. Of the PAR populations, SG and TS reached the largest sizes, KU was intermediate, and MS reached the smallest size. Of the NWS populations, GS grew to a larger size at all copper concentrations than SF, but both of these populations decreased in size with increasing Cu concentration. YC (the only

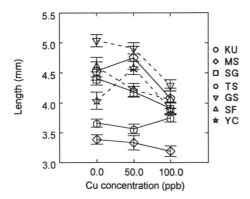

**FIG. 2.** Mean length of the different populations as a function of copper concentration. Error bars are standard errors. KU, TS, MS, and SG: PAR populations; SF, GS: NWS populations; YC: OWS population.

OWS population investigated) showed an increase in length at the intermediate Cu concentration.

### Differences Among Clones/Full-Sib Families

Significant differences in mean length (Fig. 3) among families within the NWS populations were substantial (e, f), but so were the differences among clones within PAR populations (a, b, c, and d). YC was the only population (Fig. 3g) for which there was no main effect of family. For all populations except MS there was a significant interaction between clone/full-sib family and treatment. For example, some families decreased growth linearly with increasing Cu concentration, whereas others increased in length at intermediate concentrations. Some even increased length with increasing Cu concentration. There was a significant treatment main effect only for SF and GS.

Within all the PAR populations, some clones that were not distinguishable by allozyme analysis differed in growth. Two of the clones in the SG population that were derived from the same grandmother did not differ in growth, whereas the one clone with a different allozyme pattern did differ in length compared to all other clones. In contrast, the two related TS clones (i.e., those derived from the same grandmother) were significantly different in length under control conditions, but not in the two Cu-treatments.

## Variance Components Estimates

### Variance Partitioning From ANOVA

Partitioning of variance based on clones/full-sib families within populations (Table 4) showed that residual variance was by far the largest contributor to total phenotypic variance (54.6–84.1% in parthenogenetic populations and 52.7–87.8% in sexual populations), although there were considerable differences among populations. Genetic variance (among clones) was low in three of the four PAR populations, but these were populations in which clones were indistinguishable by allozyme analysis. For the PAR population, SG, that had a high genetic variance component, two genotypes were identified by allozyme analysis. Only one of the sexual populations, SF, had a high among-families component of variance. Copper exposure contributed very little to the variance in growth for all sexual populations and to only one of the parthenogenetic populations, TS. An interaction component between copper treatment and genetic sources, however, was present in all but one of the PAR populations, MS. This interaction component was higher than the main copper treatment component in all cases. The total estimates of variance were low for three of the four parthenogenetic populations (0.38–0.69). For the remaining PAR population, SG, the level of phenotypic variance (1.37) was comparable to that found in the sexual populations (1.16–1.20 NWS and 1.57 for OWS).

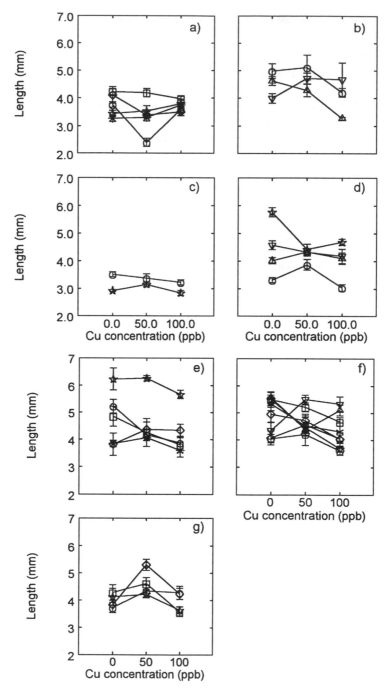

**FIG. 3.** Mean length of individual clones and families as a function of copper concentration for the different populations: (*a*) KU, (*b*) TS, (*c*) MS, (*d*) SG, (*e*) SF, (*f*) GS, and (*g*) YC. Error bars are standard errors.

**TABLE 4.** *Variance components of growth and percentages of total phenotypic variation due to different variance components for individual asexual and sexual populations\**

|  | $\sigma_G^2$ | $\sigma_E^2$ | $\sigma_{G \times E}^2$ | $\sigma_R^2$ | $\sigma_{Total}^2$ |
|---|---|---|---|---|---|
| **Parthenogenetic populations (PAR)** | | | | | |
| KU | 0.08 | 0.00 | 0.07 | 0.50 | 0.66 |
| % | 12.2 | 0.0 | 11.3 | 76.6 | |
| TS | 0.05 | 0.03 | 0.14 | 0.46 | 0.69 |
| % | 7.3 | 4.7 | 20.9 | 67.2 | |
| MS | 0.06 | 0.00 | 0.00 | 0.32 | 0.38 |
| % | 15.9 | 0.0 | 0.0 | 84.1 | |
| SG | 0.43 | 0.00 | 0.19 | 0.75 | 1.37 |
| % | 31.4 | 0.0 | 14.1 | 54.6 | |
| **New world sexual populations (NWS)** | | | | | |
| SF | 0.43 | 0.02 | 0.12 | 0.63 | 1.20 |
| % | 35.7 | 1.9 | 9.6 | 52.7 | |
| GS | 0.13 | 0.10 | 0.18 | 0.76 | 1.16 |
| % | 10.8 | 8.5 | 15.2 | 65.6 | |
| **Old world sexual populations (OWS)** | | | | | |
| YC | 0.00 | 0.09 | 0.10 | 1.38 | 1.57 |
| % | 0.0 | 5.7 | 6.4 | 87.8 | |

\* $\sigma_G^2$: "genetic" variance (among clones/families), $\sigma_E^2$: environmental variance (among Cu-treatments), $\sigma_{G \times E}^2$: genotype by environment variance, $\sigma_R^2$: residual variance (among individuals), and $\sigma_{Total}^2$: total phenotypic variance. See text for further explanation.

## Coefficients of Variation

ANOVA on within-treatment coefficients of variation showed that only the species effect was significant. Treatment, populations within species, clones/families within populations, and all the interactions were not significant. Fig. 4 shows a box plot of CV$_{Within}$ for the three species. Multiple comparisons showed that the average variabilities within single clones/full-sib families for PAR and NWS species were comparable, but that both varied less than the OWS species.

There was no difference between PAR and NWS species in across-environment coefficients of variation (Fig. 5). There were no differences among populations for

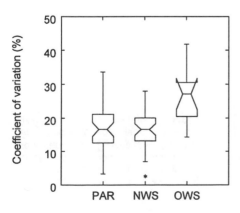

**FIG. 4.** Box plot of within treatment coefficient of variation for the three species. Notches indicate 95% confidence intervals around the median.

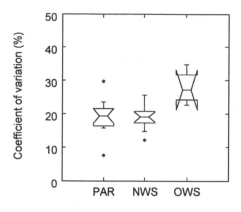

**FIG. 5.** Box plot of across treatment coefficient of variation for the three species. Notches indicate 95% confidence intervals around the median.

parthenogenetic (mean 16–21%) and NWS (mean 18–19%) populations, but $CV_{Across}$ was significantly higher (mean 27%) in the OWS population. In other words, growth of YC was more sensitive than all other populations to copper exposure.

### Relationship Between Performance and Stability

Figure 6 shows a plot of sensitivity ($s_{G \times E}^2 + s_E^2$) against population mean growth rate. There was a significant positive correlation ($r = 0.94$, $p = 0.002$) between sensitivity and the population mean, indicating that a high mean growth rate was associated with a high sensitivity to copper exposure. The plot is divided into four quadrants (see figure text) that define four combinations of growth and copper sensitivity: high sensitivity-low growth (upper left quadrant), low sensitivity-low growth (lower left), high sensitivity-high growth (upper right), and low sensitivity-high growth (low right).

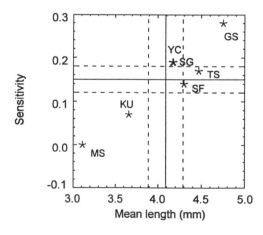

**FIG. 6.** Scatter plots of sensitivity ($s_E^2 + s_{G \times E}^2$) against performance (mean length) for the seven populations. Horizontal line: mean overall sensitivity; vertical line: mean overall length. Stippled lines mean plus/minus standard error.

Two of the PAR populations, MS and KU, are located in the low sensitivity-low growth quadrant and one of the NWS populations, GS, is in the high sensitivity-high growth quadrant. All other populations cluster around the intersection and fall within the standard error limits.

## DISCUSSION

### Average Growth

Although there were significant differences among species of *Artemia* in their response to copper (significant treatment × species interaction), the two sexual species were not more similar to each other than to the parthenogenetic species. Furthermore, differences in growth among populations were less pronounced than differences among clones and families within populations.

These results are consistent with earlier studies comparing growth and other life-history traits between sexual and parthenogenetic *Artemia* (some of which did not include exposure to toxicants). Of these, most have shown significant differences among populations within reproductive mode or species that often confound differences between sexual and parthenogenetic reproductive modes. The only study comparing growth rate of a parthenogenetic population (La Palma, France) and a NWS population (San Diego, CA) showed that parthenogenetic females grew most rapidly, sexual females were intermediate in growth rate, and males grew least (Gilchrist 1960). In a study comparing the tolerance of several parthenogenetic (including KU) and NWS (including SF) populations to copper, Browne (1980) found that parthenogenetic populations were the least tolerant, closely followed by SF. Two other NWS populations were much more tolerant (Browne 1980).

Differences in growth among clones within the parthenogenetic populations were at comparable levels to differences among sexual families. Parker (1984) compared phenotypic differences among clones of a parthenogenetic cockroach species and among sexually produced broods of its ancestral species. He found that clones from one (but not from the other two) of the parthenogenetic populations studied differed as much phenotypically from each other as did sexual broods of its ancestral species. In the present study, we generated clones by using offspring from a single parthenogenetic mother. Large differences among our generated clones would have been expected if the parthenogenetic stock populations were multiclonal. This was confirmed by allozyme analysis for only one population (SG). This population did exhibit the largest among-clone differences in growth of all parthenogenetic populations considered in our study, which is consistent with previous findings of genetic diversity and variability in life-history traits among SG clones (Browne 1988). Whereas the two SG clones sharing a common grandmother did not differ significantly in growth rate, the two TS clones derived from a single grandmother did. This suggests that in TS either maternal effects are important, that recombination occurred (and was not picked up by the allozyme analysis), or that TS is more sensitive to environmental microheterogeneity than is SG. It is worth noting that the polyploid population, KU, which is believed to be monoclonal and does not undergo recombination, also exhibited a significant difference among clones, suggesting significant maternal effects.

Significant interactions between treatment and clones/families were found in all populations but MS. Thus, particular genotypes/families did not consistently perform better or worse across a copper gradient. Parallel reaction norms for growth and reproductive traits for various environmental gradients have rarely been found among vegetative clones of plants (Sultan and Bazzaz 1993a, 1993b, 1993c). The fact that there are not specific genotypes that have higher relative fitness in all environments helps to explain the maintenance of genetic variation in natural populations (Sultan and Bazzaz 1993a; 1993b; 1993c).

The old world sexual species *Artemia urmiana* (represented by population YC in this study) is the closest relative to *Artemia parthenogenetica* (Abreu-Grobois 1987; Browne et al. 1991). This species has previously been found to be intermediate between parthenogenetic and sexual populations in reproductive performance (Browne et al. 1991). Overall it showed intermediate growth in this study, but aside from that it performed rather differently than all other populations; it was the only population for which there were no differences among families in growth rate. At the same time it was consistently the most variable population (both in terms of total phenotypic variance and residual among-individual variance). Furthermore, the YC population was the only one to show a significant increase in size at the intermediate copper concentration. Stimulation in growth rate at low toxicant concentrations (termed hormesis) has previously been observed, both in *Artemia* (Brown and Ahsanullah 1971; Jayasekara et al. 1986) and in other species (e.g., Dixon and Sprague 1981; Bodar et al. 1988; Weis et al. chapter 3), where it has been explained as the consequence of regulatory overcorrections by biosynthetic control mechanisms to low levels of inhibitory challenge (Stebbing 1982; Bodar et al. 1988).

## Variance Components Estimates

Variance among clones/families explained between 7 and 36% of the total phenotypic variance in growth for parthenogenetic and NWS populations, but there was no contribution to phenotypic variance from family in the OWS population. This could indicate a high degree of genetic fixation, possibly as a result of founder effects occurring during population bottlenecks or as a result of high levels of inbreeding in this population. In SF (NWS) and SG (PAR), variance among clones/families contributed between 2–3 times more to the total phenotypic variation than in the other NWS and PAR populations.

Soares et al. (1992) studied interclonal variation in performance of *Daphnia magna* in response to two toxicants. They reported genetic variance components for the intrinsic rate of increase, $r$, of 22% and 32% for a pesticide (3,4-dichloraniline) and sodium bromide (NaBr), respectively, which are comparable to the levels observed among clones of the parthenogenetic *Artemia* examined here. However, Soares et al.

(1992) stated that these values are low, which must be in comparison to the orders of magnitude differences in acute tolerance to pollutants among *Daphnia* clones found by Baird et al. (1990, 1991).

The within-treatment variation among individuals was similar for full-sib NWS families and clones of the parthenogenetic populations (as shown by both within-treatment CV and the residual variance component), but was higher in the OWS population. Genetic differences among siblings would have been expected to result in higher residual variation in the full-sib family design, but this was seen only for the YC population (which, on the other hand, had no genetic variance component). Previous work has shown parthenogenetic populations of *Artemia* to be more variable in some life-history traits than their sexual counterparts (Browne et al. 1984). Taken together, these studies suggest that the degree of among-individual phenotypic variance in a single clone is trait and possibly environment specific and cannot automatically be assumed to be lower within clones than within sexual populations.

The small contribution of treatment to total phenotypic variation (range: 0–8.5%) indicates that the Cu concentrations used here had relatively little effect on growth for any of the populations. The actual exposure concentrations may have been lower than expected, but attempts to measure copper concentrations with atomic absorption spectrophotometry were unsuccessful due to interference by the very high salt content of the medium. The clone/family × treatment interaction component (0–21%) was higher than the treatment component for all populations (except the parthenogenetic population, MS, that had neither a significant treatment nor interaction component). Using the combined treatment and clone/family × treatment interaction component as a measure of phenotypic plasticity placed two of the parthenogenetic populations (MS and KU) in the low sensitivity quadrant and one of the NWS populations (GS) in the high sensitivity quadrant, whereas all other populations were intermediate between these. This result was partly contradicted by the results obtained from the across-environment CV analysis, however, which indicated a significantly higher across-treatment variation in the OWS population while showing no difference between parthenogenetic and NWS populations. Similar discrepancies between different measures of phenotypic variability have previously been shown (Bierzychudek 1989).

With regard to differences in sensitivity of parthenogenetic versus sexual species to copper exposure (question 1, Introduction), we did detect significant differences among the seven *Artemia* populations in their response to copper exposure; however, the two sexual species were not more similar to each other than to the parthenogenic species. Because we found significant species × copper interaction effects, extrapolating test results from one species (e.g., *A. parthenogenetica*, which showed little influence of copper on growth in the present study) to another (e.g., *A. urmiana*, which showed its highest growth rate at intermediate copper concentrations) could be problematic. *Artemia* populations of the same species collected from geographically distinct areas responded similarly to copper exposure (question 2), suggesting that differences in conditions at source habitats did not play an important role in copper tolerance, at least after the populations had been reared in the laboratory for several generations. We found phenotypic differences not only among families within the sexual populations but also among allozymically indistinguishable clones in parthenogenetic populations.

Interactions between clone and copper treatment were a significant source of phenotypic variability for the parthenogenetic populations and accounted for up to 20% of the total phenotypic variance (question 3). However, residual variability among individuals (within-clones) accounted for the largest percent of the total phenotypic variability, ranging between 55 and 84% (question 4). This was similar for the sexual populations in which phenotypic differences among siblings explained the largest fraction of the total phenotypic variability at the population-level. These results suggest that a substantial fraction of the response variability observed among individuals in toxicant-exposed populations may be unavoidable (see also Baird and Barata, chapter 11).

Variance partitioning indicated that one of the four parthenogenetic populations tested had a total phenotypic variance between that of the two sexual species; the other three parthenogenetic populations tested ranked lower with around half the level of total phenotypic variability (however, part of this difference is due to a lowered contribution of the copper treatment to phenotypic variability in the parthenogens). Analysis of the coefficient of variation of growth rate within-treatments indicated that individuals from single clones of the parthenogenetic species differed as much from each other phenotypically, on average, as did siblings from single families of the New World sexual populations, with siblings from Old World sexual families differing somewhat more from each other. So, although our results do not indicate the employment of clones is likely to *increase* variability in toxicity test results, we saw no evidence that their use is likely to be of substantial advantage in terms of increasing the uniformity of populations' responses to toxicant exposure.

In conclusion, there are two potential problems with employing clonal organisms in ecotoxicity tests. If the responses of populations to toxicants are strongly influenced by genetic factors, then employing single clones leads to biased test results, and the degree of bias is likely to vary among chemicals in ways that are difficult to predict a priori (Baird and Barata, chapter 11). In this situation, one might consider employing a suite of clones. However, it becomes rapidly apparent that the increased effort involved in repeating every test with several clonal populations may offer no advantage over performing a single test with a genetically-mixed test population. A different problem arises if the responses of populations to toxicants are largely influenced by nongenetic factors. Then culturing single clones is not likely to provide the desired reduction in response variability. It may legitimately be decided that clonal populations in this case nevertheless provide useful test organisms if culturing them requires less effort than culturing sexual populations. Potential complications arise when the relative importance of genetic versus non-genetic factors is endpoint-dependent (e.g., if lethal responses are largely genetically determined whereas sublethal responses are much less so). There are some preliminary indications that this might be the case (Møller et al. 1996), but present understanding of these interactions is far from conclusive.

## ACKNOWLEDGMENTS

We thank Dave Jowett for statistical advice regarding the ANOVA designs, R. Dimock for image analysis advice, and G. Banta, M. Niklasson, and P. Calow for valuable discussions and for reviewing the manuscript.

## REFERENCES

Abreu-Grobois, F. A. 1987. A review of the genetics of *Artemia*. In *Artemia research and its applications. Vol. 1. Morphology, genetics, strain characterisation, toxicology*. eds. P. Sorgeloos, D. A. Bengtson, W. Declair, and E. Jaspers, 61–99. Wetteren, Belgium: Universa Press.

Abreu-Grobois, F. A., and J. A. Beardmore. 1982. Genetic differentiation and speciation in the brine shrimp *Artemia*. In *Mechanisms of speciation*, ed. C. Barigozzi, 345–376. New York: Alan R. Liss, Inc.

Baird, D. J. 1992. Predicting population response to pollution: In praise of clones. A comment on Forbes and Depledge. *Functional Ecology* 6:616–617.

Baird, D. J. 1993. Can toxicity testing contribute to ecotoxicology? *Functional Ecology* 7:510–511.

Baird, D. J., I. Barber, and P. Calow. 1990. Clonal variation in general responses of *Daphnia magna* Straus to toxic stress. I. Chronic life-history effects. *Functional Ecology* 4:399–408.

Baird, D. J., I. Barber, M. Bradley, A. M. V. M. Soares, and P. Calow. 1991. A comparative study of genotype sensitivity to acute toxic stress using clones of *Daphnia magna* Straus. *Ecotoxicology and Environmental Safety* 21:257–263.

Bierzychudek, P. 1989. Environmental sensitivity of sexual and apomictic *Antennaria*: Do apomicts have general-purpose genotypes? *Evolution* 43:1456–1466.

Bodar, C. M. V., C. J. van Leeuwen, P. A. Voogt, and D. J. Zandee. 1988. Effect of cadmium on reproductive strategy in *Daphnia magna*. *Aquatic Toxicology* 12:301–310.

Brown, B., and M. Ahsanullah. 1971. Effect of heavy metal on mortality and growth. *Marine Pollution Bulletin* 3:182–188.

Browne, R. A. 1980. Acute response versus reproductive performance in five strains of brine shrimp exposed to copper sulphate. *Marine Environmental Research* 3:185–193.

Browne, R. A. 1988. Ecological and genetic divergence of sexual and asexual brine shrimp (*Artemia*) from the Mediterranean basin. *National Geographic Research* 4:547–554.

Browne, R. A., and S. J. Bowen. 1991. Taxonomy and population genetics of *Artemia*. In *Artemia biology*, eds. R. A. Browne, P. Sorgeloos, and C. N. A. Trotman, 221–735. Boca Raton, FL: CRC Press.

Browne, R. A., S. E. Sallee, D. S. Grosch, W. O. Segreti, and S. M. Purser. 1984. Partitioning genetic and environmental components of reproduction and lifespan in *Artemia*. *Ecology* 65:949–960.

Browne, R. A., M. Li, G. Wanigasekara, S. Simonek, D. Brownlee, G. Eiband, and J. Cowan. 1991. Ecological, physiological and genetic divergence of sexual and asexual (diploid and polyploid) brine shrimp (*Artemia*). *Advances in Ecological Research* 1:41–52.

Dixon, D. G., and J. B. Sprague. 1981. Acclimation to copper by rainbow trout (*Salmo gairdneri*)—A modifying factor in toxicity. *Canadian Journal of Fisheries and Aquatic Sciences* 38:880–888.

Dutilleul, P., and C. Potvin. 1995. Among-environment heteroscedasticity and genetic autocorrelation: Implications for the study of phenotypic plasticity. *Genetics* 139:1815–1829.

Forbes, V. E., and P. Calow. 1996. Costs of living with contaminants: Implications for assessing low-level exposures. *Biological Effects of Low Level Exposures (BELLE) Newsletter* 4(3):1–8.

Forbes, V. E., and M. H. Depledge. 1993. Testing versus research in ecotoxicology: A response to Baird and Calow. *Functional Ecology* 7:509–510.

Forbes, V. E., and M. H. Depledge. 1996. Environmental stress and the distribution of traits within populations. In *Ecotoxicology: Ecological dimensions*, eds. D. J. Baird, L. Maltby, P. W. Greig-Smith, and P. E. T. Douben, 71–86. London: Chapman and Hall.

Forbes, V. E., and T. L. Forbes. 1994. *Ecotoxicology in theory and practice*. London: Chapman and Hall.

Fry, D. J. 1992. The mixed model analysis of variance applied to quantitative genetics: Biological meaning of the parameters. *Evolution* 46:540–550.

Gaddum, J. H. 1933. Reports on biological standards. III. Methods of biological assay depending on quantal response. *Medical research council, special report series no. 183*, London.

Gilchrist, B. M. 1960. Growth and form of the brine shrimp *Artemia salina* (L). *Proceedings of the Zoological Society of London* 134:221–235.

Jayasekara, S., D. B. Brown, and R. P. Sharma. 1986. Tolerance to cadmium and cadmium-binding ligands in Great Salt Lake brine shrimp (*Artemia salina*). *Ecotoxicology and Environmental Safety* 11:23–30.

Kirby, K. N. 1993. *Advanced data analysis with SYSTAT*. New York: Van Nostrand Reinhold.

Koehn, R., and B. L. Bayne. 1989. Towards a physiological and genetical understanding of the energetics of the stress response. *Biological Journal of the Linnean Society* 37:157–171.

Michaels, H. J., and F. A. Bazzaz. 1989. Individual and population responses of sexual and apomictic plants to environmental gradients. *American Naturalist* 134:190–207.

Milliken, G. A., and D. E. Johnson. 1992. *Analysis of messy data. Vol. 1. Designed experiments*. New York: Chapman & Hall.

Møller, V., V. E. Forbes, and M. H. Depledge. 1996. Population responses to acute and chronic cadmium exposure in sexual and asexual estuarine gastropods. *Ecotoxicology* 5:313–326.

OECD. 1997. Guideline for testing of chemicals, no. 202, *Daphnia* sp. acute immobilisation test and reproduction test. Paris: Organisation for Economic Cooperation and Development.

Parker, Jr., E. D. 1984. Reaction norms of development rate among diploid clones the parthenogenetic cockroach, *Pycnoscelus surinamensis*. *Evolution* 38:1186–1193.

Pilla, E., and J. A. Beardmore. 1994. Genetic and morphometric differentiation in Old World bisexual species of the brine shrimp (*Artemia*). *Heredity* 72:47–56.

Scheiner, S. M., and C. J. Goodnight. 1984. The comparison of phenotypic and genetic variation in populations of grass *Danthionia spicata*. *Evolution* 38:845–855.

Snell, T. W., and M. J. Carmona. 1995. Comparative sensitivity of sexual and asexual reproduction in the rotifer *Brachionus calyciflorus*. *Environmental Toxicology and Chemistry* 14:415–420.

Soares, A. M. V. M., D. J. Baird, and P. Calow. 1992. Interclonal variation in the performance of *Daphnia magna* Straus in chronic bioassays. *Environmental Toxicology and Chemistry* 11:1477–1483.

Sokal, R. R, and F. J. Rohlf. 1996. *Biometry*, 3rd ed. New York: WH Freeman & Co.

Stearns, S. C. 1992. *The evolution of life histories*. Oxford: Oxford University Press.

Stebbing, A. R. D. 1982. Hormesis—The stimulation of growth by low levels of inhibitors. *Science of the Total Environment* 22:213–234.

Sultan, S. E., and F. A. Bazzaz. 1993a. Phenotypic plasticity in *Polygonum persicaria*. I. Diversity and uniformity in genotypic norms of reaction to light. *Evolution* 47:1009–1031.

Sultan, S. E., and F. A. Bazzaz. 1993b. Phenotypic plasticity in *Polygonum persicaria*. II. Norms of reaction to soil moisture and the maintenance of genetic diversity. *Evolution* 47:1032–1049.

Sultan, S. E., and F. A. Bazzaz. 1993c. Phenotypic plasticity in *Polygonum persicaria*. III. The evolution of ecological breadth for nutrient environment. *Evolution* 47:1050–1731.

Sunda, W. G., and A. K. Hanson. 1987. Measurements of free cupric ion concentration in seawater by a ligand competition technique involving copper sorption onto C18 SEP-PAK cartridges. *Limnology and Oceanography* 32:537–551.

Via, S. 1984. The quantitative genetics of polyphagy in an insect herbivore. I. Genotype-environment interaction in larval performance on different host plant species. *Evolution* 38:881–895.

Weider, L. J. 1993. A test of the "general-purpose" genotype hypothesis: Differential tolerance to thermal and salinity stress among *Daphnia* clones. *Evolution* 47:965–969.

Weis, A. E., and W. L. Gorman. 1990. Measuring selection on reaction norms: an exploration of the Eurosta-Soldigo system. *Evolution* 44:820–831.

Wilkinson, L., M. A. Hill, J. P. Welna, and G. K. Birkenbeuel. 1992. *SYSTAT for Windows*, version 5 edition. Evanston, IL: Systat Inc.

Willows, R. 1994. The ecological impact of different mechanisms of chronic sub-lethal toxicity on feeding and respiratory physiology. In *Water quality and stress indicators in marine and freshwater ecosystems: Linking levels of organisation*, ed. D. W. Sutcliffe, 88–97. UK: Freshwater Biological Association.

Winer, B. J. 1971. *Statistical principles in experimental design*. New York: McGraw Hill.

Yampolsky, L. Y., and S. M. Scheiner. 1994. Developmental noise, phenotypic plasticity, and allozyme heterozygosity in *Daphnia*. *Evolution* 48:1715–1722.

*Genetics and Ecotoxicology*
Edited by V. E. Forbes
Copyright © 1999 Taylor & Francis

# 11

# Genetic Variation in the Response of *Daphnia* to Toxic Substances: Implications for Risk Assessment

Donald J. Baird and Carlos Barata

**Abstract.** *Daphnia magna* is widely used as a toxicity test species in Europe and elsewhere. Its use has been plagued with problems relating to variability in response, both within and between laboratories carrying out standard tests. Here we review the relative contribution of genetic versus environmental factors in the expression of variability in response to toxic substances in a series of studies carried out on a selected group of laboratory genotypes held within the European Union (EU) and North America. Although the general pattern emerging from this review is that environmental factors such as varying diet and culture conditions remain the major cause of interlaboratory variation in response, the consequences of genetically-based egg-size variation as a mechanism for promoting variability in test animals are described for the first time. The relationship between variability in acute and chronic response is discussed in relation to a general hypothesis that toxicity responses converge from acute-chronic, reflecting a shift from specific to general response mechanisms. Finally, the relationship between variation in laboratory stocks and variation in field populations of *D. magna* is considered in the context of extrapolation of laboratory data to the field situation.

**Keywords.** Acute toxicity; chronic toxicity; egg-size variation; extrapolation; phenotypic variation.

## INTRODUCTION

Risk assessment of new and existing chemicals requires the production of high-quality information to allow us to predict the toxicity of substances in nature. In the aquatic environment, the most commonly tested representative species is the freshwater cladoceran crustacean *Daphnia magna* (Baudo 1987). The choice of *D. magna* is no accident, as it possesses a suite of characteristics which make it an ideal candidate for laboratory culture and experimentation—small size, parthenogenetic mode of reproduction, and simple culture and dietary requirements (Baird et al. 1989a). Indeed, these characteristics make *Daphnia* a favored organism for mathematical modellers also (e.g., Gurney et al. 1990). Above all, it is its genetic uniformity which offers the greatest advantage to laboratory researchers because clonal animals such as *Daphnia* offer the prospect of fully repeatable, reproducible results. However, far from being the first animal to be certified for compliance with good laboratory practice guidelines,

*Daphnia* is capable of bringing out the worst in interlaboratory calibration exercises (Baird et al. 1989a). Causes of interlaboratory variation in toxicity tests using *Daphnia* involve a wide range of factors, including experimenter error (as evidenced by the wide deviations from a standard test protocol arising from a 1986 *Daphnia* ring test (Baird et al. 1989a)), yet perhaps the factor that has yielded the greatest speculation and controversy is the role of genetic variability among testing strains, held in different laboratories, as a source of between-test variation.

Because the ultimate aim of laboratory toxicity tests is to predict effects on organisms in the natural environment, one might perhaps question the use of clones as a step too far down the road of sacrificing test realism for laboratory convenience (see also Snell et al., chapter 9). This reflects a necessary antagonism in ecotoxicology between the need for experimental standardization (to give repeatability and reproducibility) and the need for the generation of relevant data to assess risks of long-term damage (Calow 1992). Here, using published and unpublished examples from our own work and that of others, we argue that the biology of *Daphnia*, including its parthenogenetic mode of reproduction, offers a unique tool to study the underlying nature of organism response to contaminant exposure (see also Forbes et al., chapter 10). In particular, we focus on the role clones can play in teasing apart the relative contributions of environment and genotype in the expression of a trait, in this case, the response to chemical exposure. Finally, we conclude with a consideration of how the study of *Daphnia* populations in nature might help us to improve our understanding of laboratory experiments, and how we might in future develop a more ecologically relevant approach to obtaining aquatic hazard assessment data using *Daphnia*.

## RESPONSE OF LABORATORY *DAPHNIA* CLONES TO TOXIC STRESS

The laboratory maintenance of *Daphnia* is made simple by the fact that culture systems are designed to raise females in clonal lines. Of course, removing the need for males in such cultures does not prevent their appearance, yet although males are unavoidably produced, ephippia or sexual eggs produced by fertilized females are highly visible and notoriously difficult to hatch. For these three reasons, the maintenance of genetic integrity in uniclonal laboratory cultures is straightforward. Thus, it is not surprising to note that when samples were obtained from cultures of *Daphnia magna* across the range of European testing laboratories, only one laboratory exhibited genetic heterogeneity within its cultures, and in this case, two separate clone stocks had been recently mixed together (Baird et al. 1989b). The conclusion here is that sexual reproduction is not an important source of genetic variability in laboratory stocks. This particular study arose following the results of a 1986 interlaboratory ring test (Cabridenc 1986), which aimed to test the effectiveness of a *D. magna* chronic test protocol. The uninspiring results of the ring test, which examined the effect of sodium bromide and 3,4-dichloraniline on reproduction, revealed the maximum possible variation in results between laboratories. No observed effect concentrations, or NOECs (defined as

**TABLE 1.** Daphnia magna *clones discussed in this chapter and their origins**

| Code | Country of origin | Source | [1] | [2] | [3] | [4] | [5] | [6] | [7] | [8] | [9] |
|---|---|---|---|---|---|---|---|---|---|---|---|
| clone A | France | testing lab | X | X | X | X | X | X | X | X | X |
| clone A-1 | Netherlands | testing lab | X | X | X | X | X | | | | |
| clone A-2 | UK/France | testing lab | X | X | | | | | | | |
| clone B | Germany | testing lab | X | X | X | X | X | | | | |
| clone C | Italy | testing lab | X | X | X | | X | X | X | X | X |
| clone D | Belgium | testing lab | X | X | | X | | | | | |
| clone E | Norway | testing lab | X | X | X | X | X | | | | |
| clone F | Canada | testing lab | X | X | X | X | X | X | X | X | X |
| clone S-1 | UK | pet shop | X | X | X | X | X | X | X | X | X |
| clone S-2 | UK | local pond | X | X | X | | X | | | | |

*[1] Baird et al. 1989b; [2] Soares 1989; [3] Baird et al. 1990; [4] Baird et al. 1991; [5] Soares et al. 1992; [6] Baird and Barata 1997; [7] Barata and Baird 1998; [8] Barata et al.1998; [9] all references to unpublished work by Baird and Barata. *N.B.* the four genotypes used in the work after 1992 have been repeatedly confirmed by electrophoresis to possess the same allozyme patterns as described in Baird et al. (1991). Whether the clone genomes have been conserved over this period (>100 generations) is unknown, but clonal divergence through accumulation of mutations cannot be discounted. See also discussion in the text.

the concentration within an exposure series immediately below the lowest concentration that differs significantly from the control in some toxicity response, in this case, reduction in viable egg production) ranged from below the lowest test concentration to above the highest (Baird et al. 1989b). Although in fact this complete failure of the ring test to achieve its objectives was largely a consequence of misinterpretation of the test protocol, that resulted in environmental variation (in food level), it nevertheless stimulated interest in the possibility that genetic factors could play a part in interlaboratory test variation. Further studies carried out on a smaller subset of these EC ring test clones, which included genotypes obtained from within and outside Europe, revealed that although interclonal variation clearly existed, the implications of the findings for toxicity testing were more positive than might have been expected (Table 1).

In a definitive series of experiments, laboratory clones of *D. magna* were maintained for >30 generations under highly controlled laboratory conditions and their offspring tested for their lethal and sublethal responses to a range of toxic substances using standard test protocols (acute 48 hour lethal test; 21-day chronic sublethal test of reproductive inhibition; OECD 1997 gives the most recently updated methods). Baird et al. (1990, 1991) studied variation in acute lethal response ($LC_{50}$) among different clones of *D. magna* and noted that although significant differences in response to all chemicals existed among clones, in all cases these were within an order of magnitude (Table 2), with only cadmium being a notable exception (see below). However, there was also heterogeneity in the degree of within-clone variation, both within and between the compounds tested. However, this within-clone variability showed no apparent concordance among clones, i.e., clones could not be distinguished as "more variable" or "less variable" in their responses across different compounds.

A linked study (Soares et al. 1992) on the response of the same six genotypes, plus three others (see Table 1), to chronic, sublethal exposure to the two substances

**TABLE 2.** *Interclonal variation in response to toxic substances as measured using a standard 48 hour acute test with <24 hour old neonate* Daphnia magna[†]

| Substance | n | Range (ppb) | Source |
|---|---|---|---|
| cadmium | 8 | 0.6–116* | Baird et al. 1990 |
|  | 6 | 4–116 | Baird et al. 1991 |
|  | 4 | 24–233 | Barata et al. 1998 |
| zinc | 6 | 755–1831 | Baird et al. 1991 |
|  | 4 | 601–962 | Barata et al. 1998 |
| copper | 6 | 11–41 | Baird et al. 1991 |
|  | 4 | 22–112 | Barata et al. 1998 |
| uranium | 4 | $17 \times 10^3$–$25 \times 10^3$ | Barata et al. 1998 |
| manganese | 6 | $5 \times 10^3$–$56 \times 10^3$ | Baird et al. 1991 |
| chromium | 6 | 100–288 | Baird et al. 1991 |
| sodium bromide | 9 | $6 \times 10^6$–$8 \times 10^6$* | Soares 1989 |
|  | 6 | $7 \times 10^6$–$9 \times 10^6$ | Baird et al. 1991 |
| dodecyl benzyl sulphonate | 6 | $11 \times 10^3$–$23 \times 10^3$ | Baird et al. 1991 |
| linear alkyl sulphonate | 6 | $8 \times 10^3$–$15 \times 10^3$ | Baird et al. 1991 |
| 3,4-dichloroaniline | 9 | 79–>500* | Soares 1989 |
|  | 8 | 66–501* | Baird et al. 1990 |
|  | 6 | 628–2253 | Baird et al. 1991 |

[†]*n* indicates the number of clones tested. All data are for animals from bulk cultures, except those indicated as*, which were obtained from individual cultures. All data are based on actual measured concentrations.

employed in the 1986 ring test again revealed significant differences in response. Again variation was within an order of magnitude, and if anything, chronic responses were less variable than acute responses to the same chemical. Moreover, in no study was significant concordance in response to chemical exposure within the clone groups recorded for acute or chronic exposure, indicating that no generally tolerant or generally intolerant genotypes existed even within classes of substance such as metals. However, this interpretation must be treated with caution, first, because only a small number of genotypes was tested, genotypes which were raised in the highly canalized, static environment of the laboratory. Second, because for most substances, variation was within an order of magnitude, and recent studies (Baird and Barata 1997; see also below) have indicated that clone responses will vary over time within a similar range.

For one substance, the heavy metal cadmium, the range in response was remarkable—over three orders of magnitude within one group of eight clones tested (Baird et al. 1990; see Tables 1 and 2) in comparison with less than an order of magnitude for the other substances studied (Fig. 1).

This result was significant for a number of reasons. First, it demonstrated that chance variation in response among genotypes could be significant from the point of view of risk assessment, where such a degree of variation in response could strongly influence the predicted risk of exposure to a given substance, which is defined here as the ratio of predicted environmental concentration to perceived hazardous concentration. If the choice of a particular genotype could result in variation in concentration-response of up to three orders of magnitude, then the current safety margin approach, based on arbitrary multiplicative safety factors, would be seriously undermined.

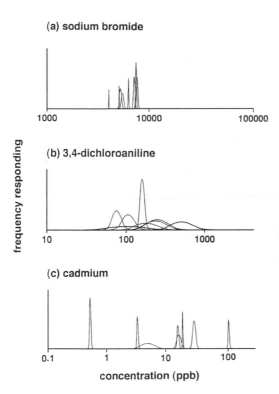

**(a) sodium bromide**

**(b) 3,4-dichloroaniline**

**(c) cadmium**

frequency responding

concentration (ppb)

**FIG. 1.** Probability density functions for the LC$_{50}$ of eight genotypes (corresponding to those used in Baird et al. 1990) exposed to (*a*) sodium bromide, (*b*) 3,4-dichloroaniline, and (*c*) cadmium. Decreasing kurtosis in the distributions indicates greater intraclonal variability in response.

Second, it raised the question: why was cadmium response more variable than for the other substances tested? Noting that cadmium is a nonessential metal, it could be argued that resistance to essential metal exposure is based on preadapted biochemical machinery. Because effective metabolism of essential metals is ubiquitous and presumably under continuous directional selection, animals with poor ability to regulate internal metal levels would be continuously weeded out of all natural populations. In contrast, regulation of nonessential metals is important only where such metals occur at high levels. Whether a clone was derived from such a population, and thus was cadmium resistant, would therefore be a highly contingent phenomenon and potentially subject to extreme variation. This hypothesis is testable; however, more recent research in our laboratory, still ongoing at the time of writing, has so far failed to support it (Barata et al. 1998). In an experimental study of a subset of four of the extreme genotypes in terms of cadmium sensitivity (clones A, C, F, and S-1 from Table 1), animals were exposed to four metals: two nonessential (cadmium and uranium) and two essential (zinc and copper). Copper and cadmium were the most toxic metals, and both exhibited the greatest range in response in comparison to uranium and zinc (Table 3). Interestingly, although the four genotypes tested in this later study included the highest (clone S-1) and lowest (clone C) sensitivity genotypes from the earlier study (Baird et al. 1990), the range in response to cadmium was much reduced,

**TABLE 3.** *The acute response of four genotypes of Daphnia magna to metal stress in waters of different hardness**

| Genotype | Intermediate hardness (90 mg L$^{-1}$ as CaCO$_3$) | | | | High hardness (180 mg L$^{-1}$ as CaCO$_3$) | | | |
|---|---|---|---|---|---|---|---|---|
| | Cd$^{2+}$ (ppb) | Zn$^{2+}$ (ppb) | UO$^{2-}$ (ppm) | Cu$^{2+}$ (ppb) | Cd$^{2+}$ (ppb) | Zn$^{2+}$ (ppb) | UO$^{2-}$ (ppm) | Cu$^{2+}$ (ppb) |
| C | 23 (12–40) | 1063 (922–1222) | 10 (7–16) | 15 (13–19) | 86 (72–113) | 962 (797–1160) | 25 (16–27) | 112 (87–131) |
| A | 10 (5–14) | 457 (331–589) | 6 (4–8) | 15 (13–19) | 24 (10–26) | 601 (504–736) | 22 (12–36) | 22 (13–26) |
| F | 77 (53–93) | 397 (208–521) | 6 (4–8) | 3 (2–4) | 48 (25–94) | 785 (521–846) | 17 (10–25) | 23 (19–30) |
| S-1 | 106 (95–127) | 678 (583–748) | 10 (6–16) | 5 (3–9) | 233 (174–279) | 623 (555–799) | 20 (11–29) | 29 (17–51) |

*Values are the 48 hour LC$_{50}$±95% confidence limits for the free ion concentrations, which were calculated using a geochemical speciation code (see Barata et al. 1998 for further details).

spanning one rather than three orders of magnitude. This failure to obtain similar results to the previous study indicates a problem in long-term repeatability of results. However, it should be noted that in the first series of experiments (Baird et al. 1990), repeat estimates were obtained to ensure the high and low sensitivities were not in error, and in all cases, results were repeatable, within a 20% range in concentrations for each clone. Explanations for this lack of constancy in clone performance over time include mutation within clonal lines (over 100 generations of each clone have elapsed since the original experiments) and changes in the culture medium in terms of medium composition and food supply. However, that this might be due to accumulation of mutations within cultures seems less likely because genetic drift tends to eliminate such mutations in bulk culture (Toline and Lynch, 1994) and neither the culture medium used nor the food supply had changed appreciably. One further possibility is that cultures could have been subject to occasional stress from diseases or other sudden unavoidable changes in conditions (e.g., constant room temperature failure). This type of rapid environmental shock could conceivably result in novel gene expression

### (a) copper

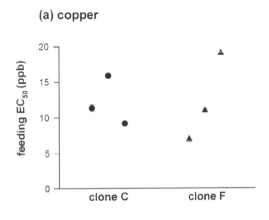

### (b) cadmium

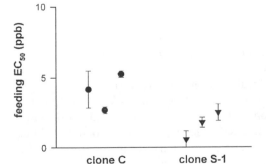

**FIG. 2.** Intraclonal variation in non-lethal response to (*a*) copper and (*b*) cadmium stress in *Daphnia magna*. The $EC_{50}$ indicates a 50% suppression of ingestion rate, measured on groups of juvenile animals over 24 hours. Replicate trials were conducted over a period of three months. Differences within clones were significant for copper exposure, $p < 0.05$, but not for cadmium exposure, $p > 0.05$ (Barata and Baird, unpublished data).

without a change in individual clonal genomes, as has been reported for insects subject to pesticide stress (Scott 1995; Taylor and Feyereisen 1996), but without further evidence, this remains speculation. In a more recent study on constancy in clone performance, again with the same group of genotypes, Baird and Barata (1997) found that even over relatively short periods of less than 10 generations under constant conditions, within-clone responses, measured as cadmium $LC_{50}$, varied by over an order of magnitude. In more recent studies (Barata and Baird, unpublished data) on a nonlethal response to stress (feeding inhibition) in this same group of four clones, there was significant fluctuation in response over time within a genotype for one substance tested, copper, but no significant fluctuation within a genotype for cadmium, the other substance studied (Fig. 2).

These results again reinforce the conclusion that variability is largely uncontrollable and may be influenced by a host of conditions, including disease incidence, water quality factors, and micro-environmental fluctuations (see below and Forbes et al., chapter 10).

## INTERCLONAL VARIATION IN OFFSPRING SIZE: A SOURCE OF UNCONTROLLABLE VARIATION?

Given that clone response to a toxic substance such as cadmium can fluctuate from generation to generation, largely ruling out stress-related novel gene expression (which may play a part over more long-term fluctuations), can we give any other explanation for this phenomenon? Enserink et al. (1990) described how changes in ration level may result in offspring size variation in *D. magna*, which can lead to three-fold differences in sensitivity. More recently, Baird and Barata (1997) and Barata and Baird (1998) have described how interclonal differences in threshold size at maturity in *D. magna* resulted in differences in the range of egg sizes within a clutch and, therefore, in sensitivity to cadmium, again using the same clone group (A, C, F, and S-1). They describe how slight, uncontrollable variations in micro-environment that arise within culture vessels due to local food depletion can translate into major differences in maturation time, leading in turn to the production of young of different size. The probability of late maturation did vary among the four clones tested: in the culture system used, late maturing, larger animals tended to produce larger young than early maturing, smaller animals. The mechanism producing this variation was found to be both subtle and complex: in *Daphnia*, egg size is a function of food availability, and in fixed-volume cultures, animals from small eggs delay maturity to the sixth instar, as they do not reach the maturity size threshold in the fifth instar. As a result, they are larger at maturity and more likely to locally deplete their food supply (resulting in transient food shortages). These transient food shortages tend to favour the production of larger offspring. Of course the opposite is true for animals hatching from large eggs. Unpublished data from these experiments showing egg size variation within a clutch (Fig. 3) indicates that although egg size distribution tends to be highly skewed, differences in the median egg size clearly exist among the clones tested at both high and low ration levels.

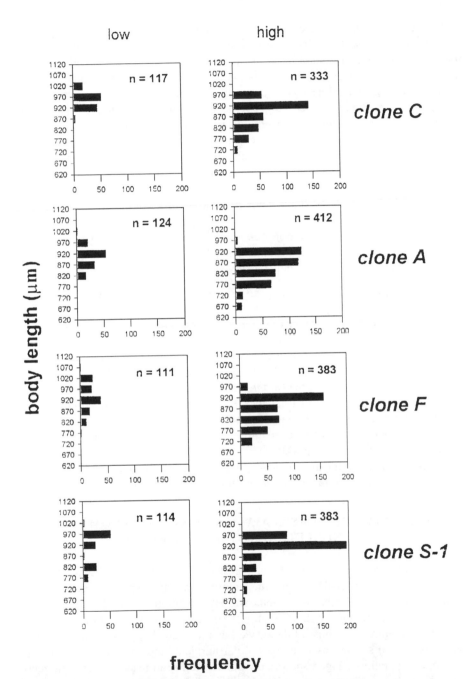

**FIG. 3.** Variability in newborn length among four laboratory clones of *Daphnia magna* Straus at low (0.4 mgC l$^{-1}$) and high (1.8 mgC l$^{-1}$) food. Data are unpublished, but culture conditions were identical to those described in Barata and Baird (1998). Newborn length was estimated from egg diameter using the relationship of Lampert's Fig. 1 (1993).

Although egg size will be strongly influenced by maternal food level, the egg-size response to maternal food level is itself genetically variable. However, the practical message is that irrespective of the clone used, egg size, and hence sensitivity of the neonates, will vary significantly in relation to micro-environmental fluctuations in maternal food supply, which cannot be easily controlled. For some genotypes, particularly clone A (the recommended genotype, as specified in the latest Organization for Economic Co-operation and Development guidelines (OECD 1997)), the range in egg size exhibited under standard culture conditions leads to offspring that fall on both sides of the maturity threshold. This results in a bimodal distribution of maturation time within clutches of genetically identical siblings (for a detailed explanation of this, see Barata and Baird 1998). As a consequence, acute sensitivity of neonates is unpredictable within certain limits, perhaps up to an order of magnitude (Baird and Barata 1997).

Given these findings, which are of course based on an extremely small, yet important sample of *D. magna* genotypes routinely used for chemical screening within the European Union, we can draw some tentative conclusions:

1. Acute sensitivity to toxic stress appears to be more genetically variable than chronic sensitivity.
2. There are no generally tolerant or intolerant genotypes to chemical exposure, even within similar classes of substance.
3. The response of individual genotypes to a given chemical will vary over time, by as much as an order of magnitude or more, for both acute and chronic exposures.
4. Further standardization of the *Daphnia* acute and chronic tests would seem to be a futile effort, given the stochastic nature of the response variability. In particular, the choice of clone A as the standard genotype (because it was the most commonly cultured clone in Europe) has been unfortunate, as it is the most prone to stochastic variation of all the genotypes studied.
5. Environmental modification of genotype response is still poorly understood, particularly in relation to the role of environmental fluctuations on gene expression.

## DO GENOTYPE RESPONSES ALWAYS CONVERGE FROM LETHAL TO NONLETHAL EXPOSURE LEVELS?

A previous study (Soares et al. 1992) suggested that for the two substances tested, sodium bromide and 3,4-dichloroaniline, among-genotype responses were less variable under chronic stress than under acute stress, as mentioned above; a similar pattern was found for cadmium by Baird et al. (1990). In a recent study in our laboratory (Barata and Baird, unpublished data), we have studied this phenomenon in more detail using two substances with specific yet unrelated modes of action: cadmium and methyl parathion. Fig. 4 shows how the variability in response between genotypes changes if we compare a nonlethal response of increasing intensity (expressed as different levels of proportional inhibition of feeding) with a lethal response. In the case of cadmium, we can clearly see that variability in response increases

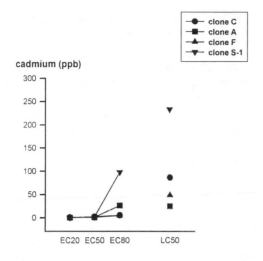

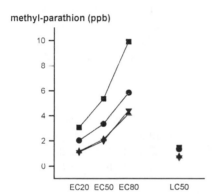

cadmium (ppb)

methyl-parathion (ppb)

**FIG. 4.** Responses of four laboratory clones expressed as proportional effects on the ingestion rate of groups of juvenile animals ($EC_{20}$, $EC_{50}$, and $EC_{80}$, measured over 24 hours) and mortality ($LC_{50}$, measured over 48 hours) when exposed to methyl parathion and cadmium.

continuously as we move from low nonlethal to lethal stress levels. In contrast, the opposite is observed for methyl parathion, where variability in response among the genotypes is greater for the nonlethal response than for the lethal response. It should be noted that for cadmium, the lethal response occurs at higher concentrations than the nonlethal response, yet for methyl parathion, feeding only becomes inhibited at supralethal concentrations (as measured over short time periods). We must make a distinction here between nonlethal exposure levels and nonlethal responses; as noted by Allen et al. (1995), feeding is only a toxicologically important endpoint where it becomes affected at sublethal levels. These results fail to either clearly support or reject the hypothesis of convergence, and their significance is not yet fully understood. More substances need to be studied, and with a wider range of genotypes in order to see if clear patterns in lethal-nonlethal exposure continue to emerge.

## WILD VERSUS LABORATORY POPULATIONS OF *DAPHNIA*: ARE THEY REALLY DIFFERENT?

In the wild, *D. magna* occupies small, eutrophic pond habitats, in contrast with other cladocerans that tend to live in open water, oligotrophic habitats. Genetic studies on *D. magna* suggest that in the wild, genotypes tend to exhibit strong phenotypic plasticity in a range of life-history traits (e.g., Young 1979), and as a result are strikingly similar in their responses over a wide range of environmental conditions. This is consistent with the general view that parthenogenetic species tend to produce general-purpose genotypes that are stress-tolerant in terms of resistance to changes in the abiotic environment (cf. Forbes et al., chapter 10).

When we establish a laboratory population, we select for a certain kind of phenotype. This is clear from life-history comparisons among clones (e.g., Bradley et al. 1993). This bottleneck selection favors genotypes with early reproduction, tolerance to constant temperature, photoperiod, and high food levels. It has been suggested (Baird 1992) that we may be unable to predict the breadth of response of wild populations by using lab-cultured strains. After all, we establish laboratory cultures by selecting out individuals who are intolerant to laboratory conditions, and who are by inference less tolerant to other forms of stress. However, given that field populations consist of large numbers of phenotypically similar genotypes, this may be less problematic than was first thought, since it may be that laboratory and field populations share similar characteristics of phenotypic variance. However, this has not been studied, and we still know little about the breadth in response of natural populations of *Daphnia* in comparison with that which we know about a relatively small yet intensively-studied group of laboratory genotypes. This linkage between field and laboratory population responses remains a central challenge of the research effort to validate the central assumptions of risk assessment, and is an important avenue for future research (see also Snell et al., chapter 9).

## CONCLUSIONS

Evidence from a range of studies supports the view that for most substances, genetic variation within laboratory strains of *Daphnia* is not likely to be an important source of variation in laboratory toxicity test results. Of course, this may reflect the fact that only a small number of genotypes has been studied, and in those studies, only a few substances have been screened. However, the observation that acute response to cadmium can vary by three orders of magnitude illustrates that genetic variation may yet prove important for particular substances, and we should be vigilant. Although hypotheses have been forwarded that attempt to explain why some substances reveal greater interclonal variation in response than others, none of these have yet been verified, although the general picture emerging is that the degree of genetic variation in response seems unpredictable and substance-specific. This most probably reflects the propensity for variation in the defence mechanisms used in exposure avoidance and detoxification rather than in the mode of action of the toxic substance per se. For this reason, it is arguable that contingency, chance variations in molecular-biochemical makeup between genotypes, or historical preadaptation, for example, will play an important role in determining the response of natural populations to exposure to toxic substances.

Given that variability in response to a given substance does not seem to be predictable, what are the implications for the use of *Daphnia* data in hazard assessment? It is our view that this finding is probably not unique to *Daphnia*, despite its clonal mode of reproduction and high levels of phenotypic plasticity. Indeed, it could be argued that if anything, laboratory-held *Daphnia* are likely to be less variable in their response than other sexually reproducing organisms (but see also Forbes and Depledge 1996 for an alternative view). For this reason, we believe that the inherent unpredictability in response among genotypes to specific substances emphasizes the need to maintain or even increase safety margins, in terms of the so-called multiplicative "application factors" used in formal risk assessment. Given that one substance out of the small number tested has yielded a 1000-fold difference in response among nine phenotypically similar genotypes, is this indicative of perhaps even greater variability in response hidden in natural populations? Of course, it is understood that the release of toxic substances into the environment cannot take place without the loss of some individuals from an exposed group of organisms. This risk has recently been quantified in terms of an assumption of 95% protection of all species, based on a lognormal tolerance distribution (Wagner and Løkke 1991; but see also Hopkin 1993). A similar approach might well be applied to the distribution of genotype sensitivities within a population or a species. Although this seems a reasonable first step toward quantifying risks to natural populations, it makes an important yet presently untested assumption: that the laboratory tests used to derive the tolerance distribution will themselves yield tolerance estimates which are randomly distributed with respect to the natural population distribution. As has been previously pointed out (Baird 1992), there is much evidence to suggest that laboratory estimates are highly skewed toward the more tolerant end of the population tolerance distribution, thus suggesting that current risk assessment methods may be less protective than we think.

The ultimate aim of risk assessment should be to provide us with information concerning the risks to long-term survival of populations: among other things, this will be dependent on the maintenance of a minimum genome size. In short-term pollution incidents involving transient pulses of high concentrations of toxic substance, we might expect the loss of extreme genotypes from the population. Given the results obtained from *Daphnia*, we might expect that short-term pollution (pulse) events will present a greater risk to populations in terms of a threat to minimum viable genome size than background contamination (press) exposure. If we are to begin to incorporate genetic structure into our ecotoxicological approaches, then this would be a good place to start.

## ACKNOWLEDGMENTS

Thanks as ever go to those who enjoyed the "*Daphnia* experience" at Sheffield and Stirling (you know who you are) and to Valery Forbes for her patience during the slow gestation of this chapter. Carlos Barata thanks EC DGXII for support through the TMR Fellowship scheme.

## REFERENCES

Allen, Y., P. Calow, and D. J. Baird. 1995. A mechanistic model of contaminant-induced feeding inhibition in *Daphnia magna*. *Environmental Toxicology and Chemistry* 14:1625–1630.

Baird, D. J. 1992. Predicting population response to pollutants: A reply to Forbes and Depledge. *Functional Ecology* 6:616–619.

Baird, D. J., I. Barber, M. C. Bradley, P. Calow, A. Girling, and A. M. V. M. Soares. 1989a. The long-term maintenance of *Daphnia magna* Straus for use in ecotoxicity tests: Problems and prospects. In *Proceedings of the first European conference on ecotoxicology*, eds. H. Løkke, H. Tyle, and F. Bro-Rasmussen, 144–148. Lyngby, Denmark.

Baird, D. J., I. Barber, M. C. Bradley, P. Calow, and A. M. V. M. Soares. 1989b. The *Daphnia* bioassay: A critique. *Hydrobiologia* 188/189:403–406.

Baird, D. J., I. Barber, and P. Calow. 1990. Clonal variation in general responses of *Daphnia magna* to toxic stress I: Chronic life-history effects. *Functional Ecology* 4:399–407.

Baird, D. J., I. Barber, M. C. Bradley, A. M. V. M. Soares, and P. Calow. 1991. A comparative study of genotype sensitivity to acute toxic stress using clones of *Daphnia magna* Straus. *Ecotoxicology and Environmental Safety* 21:257–265.

Baird, D. J., and C. Barata. 1997. Variability in the response of *Daphnia* clones to toxic substances: Are safety margins being compromised? *Archives of Toxicology* Suppl. 20:399–406.

Barata, C., and D. J. Baird. 1998. Phenotypic plasticity and constancy in life-history traits in laboratory clones of *Daphnia magna* Straus: Effects of neonatal length. *Functional Ecology* 12:442–452.

Barata, C., D. J. Baird, and S. J. Markich. 1998. Influence of genetic and environmental factors on the tolerance of *Daphnia magna* straus to essential and non-essential metals. *Aquatic Toxicology* 42:115–137.

Baudo, R. 1987. Ecotoxicological testing with *Daphnia magna*. *Memorie dell'Istituto Italiano di Idrobiologia* 45:509–518.

Bradley, M. C., C. Naylor, D. J. Baird, I. Barber, A. Soares, and P. Calow. 1993. Reducing variability in *Daphnia* toxicity tests: A case for further standardization. In *Progress in the standardization of aquatic bioassays*, eds. A. Soares and P. Calow, 57–70. Boca Raton, FL: Lewis Publishers.

Cabridenc, R. 1986. Exercice d'intercalibration concernant une méthode de détermination de l'ecotoxicité à moyen terme des substances chimiques vis-à-vis des daphnies. *Unpublished Report to the European Commission*, 201. Contract W/63/476 (214)Ref. I.R.C.H.A.D. 8523. Vert-le-Petit, France.

Calow, P. 1992. The three Rs of ecotoxicology. *Functional Ecology* 6:617–619.

Enserink, L., W. Luttmer, and H. Maas-Diepeveen. 1990. Reproductive strategy of *Daphnia* affects the sensitivity of its progeny in acute toxicity tests. *Aquatic Toxicology* 17:15–25.

Forbes, V. E., and M. H. Depledge, 1996. Environmental stress and the distribution of traits within populations. In *Ecotoxicology: Ecological dimensions*, eds. D. J. Baird, L. Maltby, P. W. Greig-Smith, and P. E. T. Douben, 71–86. London: Chapman and Hall.

Gurney, W. S. C., E. McCauley, R. M. Nisbet, and W. W. Murdoch. 1990. The physiological ecology of *Daphnia*: A dynamic model of growth and reproduction. *Ecology* 71:716–732.

Hopkin, S. 1993. Ecological implications of '95% protection levels' for metals in soil. *Oikos* 66:137–141.

Lampert, W. 1993. Phenotypic plasticity of the size at first reproduction in *Daphnia*: The importance of maternal size. *Ecology* 74:1455–1466.

OECD 1997. Guideline for Testing of Chemicals no. 202. *Daphnia* sp. acute immobilisation and reproduction test. Paris: Organization for Economic Cooperation and Development.

Scott, J. 1995. The molecular genetics of resistance: Resistance as a response to stress. *Florida Entomologist* 78:399–414.

Soares, A. M. V. M. 1989. Clonal variation in life-history traits in *Daphnia magna* Straus (Crustacea: Cladocera). Implications for ecotoxicology. Unpublished Ph.D. thesis. Sheffield, UK: University of Sheffield.

Soares, A. M. V. M., D. J. Baird, and P. Calow. 1992. Interclonal variation in the performance of *Daphnia magna* Straus in chronic bioassays. *Environmental Toxicology and Chemistry* 11:1477–1483.

Taylor, M., and R. Feyereisen. 1996. The molecular biology and evolution of resistance to toxicants. *Molecular Biology and Evolution* 13:719–734.

Toline, C. A., and M. Lynch. 1994. Mutational divergence of life-history traits in an obligate parthenogen. *Genome* 37:33–35.

Wagner, C., and H. Løkke. 1991. Estimation of ecotoxicological protection levels from NOEC toxicity data. *Water Research* 25:1237–1242.

Young, J. P. W. 1979. Enzyme polymorphism and cyclic parthenogenesis in *Daphnia magna*. I. Selection and clonal diversity. *Genetics* 92:953–970.

# Index